BIOLOGICAL AND MEDICAL PHYSICS, BIOMEDICAL ENGINEERING

BIOLOGICAL AND MEDICAL PHYSICS, BIOMEDICAL ENGINEERING

The fields of biological and medical physics and biomedical engineering are broad, multidisciplinary and dynamic. They lie at the crossroads of frontier research in physics, biology, chemistry, and medicine. The Biological and Medical Physics, Biomedical Engineering Series is intended to be comprehensive, covering a broad range of topics important to the study of the physical, chemical and biological sciences. Its goal is to provide scientists and engineers with textbooks, monographs, and reference works to address the growing need for information.

Books in the series emphasize established and emergent areas of science includingmolecular, membrane, and mathematical biophysics; photosynthetic energy harvesting and conversion; information processing; physical principles of genetics; sensory communications; automata networks, neural networks, and cellular automata. Equally important will be coverage of applied aspects of biological and medical physics and biomedical engineering such as molecular electronic components and devices, biosensors, medicine, imaging, physical principles of renewable energy production, advanced prostheses, and environmental control and engineering.

Bharat Bhushan

Biophysics of Human Hair

Structural, Nanomechanical, and Nanotribological Studies

With 94 Figures

Springer

Professor Bharat Bhushan
Ohio State University
Nanoprobe Laboratory for Bio- and
Nanotechnology and Biomimetics (NLB^2)
201 West 19th Avenue
Columbus, OH 43210-1142, USA
bhushan.2@osu.edu

Biological and Medical Physics, Biomedical Engineering ISSN 1618-7210

ISBN 978-3-642-15900-8 e-ISBN 978-3-642-15901-5
DOI 10.1007/978-3-642-15901-5
Springer Heidelberg Dordrecht London New York

Library of Congress Control Number: 2010938615

Cover design: Integra Software Services Pvt. Ltd., Pondicherry

Printed on acid-free paper

Springer is part of Springer Science+Business Media (www.springer.com)

Preface

Human hair is a nanocomposite biological fiber. Maintaining the health, feel, shine, color, softness, and overall aesthetics of the hair is highly desired. Hair care products such as shampoos and conditioners, along with damaging processes such as chemical dyeing and permanent wave treatments, affect the maintenance and grooming process and are important to study because they alter many hair properties. Nanoscale characterization of the cellular structure, mechanical properties, and morphological, frictional, and adhesive properties (tribological properties) of hair are essential to evaluate and develop better cosmetic products and to advance the understanding of biological and cosmetic science. The atomic/friction force microscope (AFM/FFM) and nanoindenter have become important tools for studying the micro/nanoscale properties of human hair. In this book, we present a comprehensive review of structural, nanomechanical, and nanotribological properties of various hair and skin as a function of ethnicity, damage, conditioning treatment, and various environments. Various cellular structures of human hair and fine sublamellar structures of the cuticle are identified and studied. Nanomechanical properties such as hardness, elastic modulus, tensile deformation, fatigue, creep, and scratch resistance are discussed. Nanotribological properties such as roughness, friction, and adhesion are presented, as well as investigations of conditioner distribution, thickness, and binding interactions. To understand the electrostatic charge buildup on hair, surface potential studies are also presented.

The book should serve as a reference book on the biophysics of human hair and hair treatments.

The research reported in this book was supported by Procter & Gamble Co. in Cincinnati, Ohio, and Kobe, Japan. I would like to thank Carmen LaTorre who had assisted in various publications on hair research. I would also like to thank my wife Sudha, who has been forbearing during the preparation of this book.

Powell, OH
November 2010

Bharat Bhushan

Contents

Chapter 1
Introduction—Human Hair, Skin, and Hair Care Products

This book presents the biophysics of human hair and hair treatment. It deals with the structure of hair, the nanomechanical characterization, nanotribological characterization, the conditioner thickness distribution and binding interactions on hair surface, and surface potential studies. It is the first book on the biophysical properties of hair.

1.1 Human Hair

Figure 1.1a shows a schematic of a human hair fiber with its various layers of cellular structure (Feughelman, 1997; Negri et al., 1993; Robbins, 1994; Zviak, 1986; Jollès et al., 1997; Smith and Swift, 2002). Hair fibers (about 50–100 μm in diameter) consist of the cuticle and cortex, and in some cases medulla in the central region. All are composed of dead cells which are mainly filled with keratin protein. Table 1.1 displays a summary of the chemical species of hair (Chen and Bhushan, 2005). Depending on its moisture content, human hair consists of approximately 65–95% keratin proteins, and the remaining constituents are water, lipids (structural and free), pigment, and trace elements. Proteins are made up of long chains of various mixtures of some 20 or 50 amino acids. Each chain takes up a helical or coiled form. Among numerous amino acids in human hair, cystine is one of the most important amino acids. Every cystine unit contains two cysteine amino acids in different chains which lie near to each other and are linked together by two sulfur atoms, forming a very strong bond known as a disulfide linkage; see Fig. 1.1b (Gray, 2003). In addition to disulfide bonds, hair is also rich in peptide bonds, and the abundant CO- and NH-groups present give rise to hydrogen bonds between groups of neighboring chain molecules. The distinct cystine content of various cellular structures of human hair results in a significant effect on their physical properties. A high cystine content corresponds to rich disulfide cross-links, leading to high mechanical properties. The species responsible for color in hair is the pigment melanin, which is located in the cortex of the hair in granular form.

An average head contains over 100,000 hair follicles, which are the cavities in the skin surface from which hair fibers grow. Each follicle grows about 20 new hair

B. Bhushan, *Biophysics of Human Hair*, Biological and Medical Physics, Biomedical Engineering, DOI 10.1007/978-3-642-15901-5_1,

a)

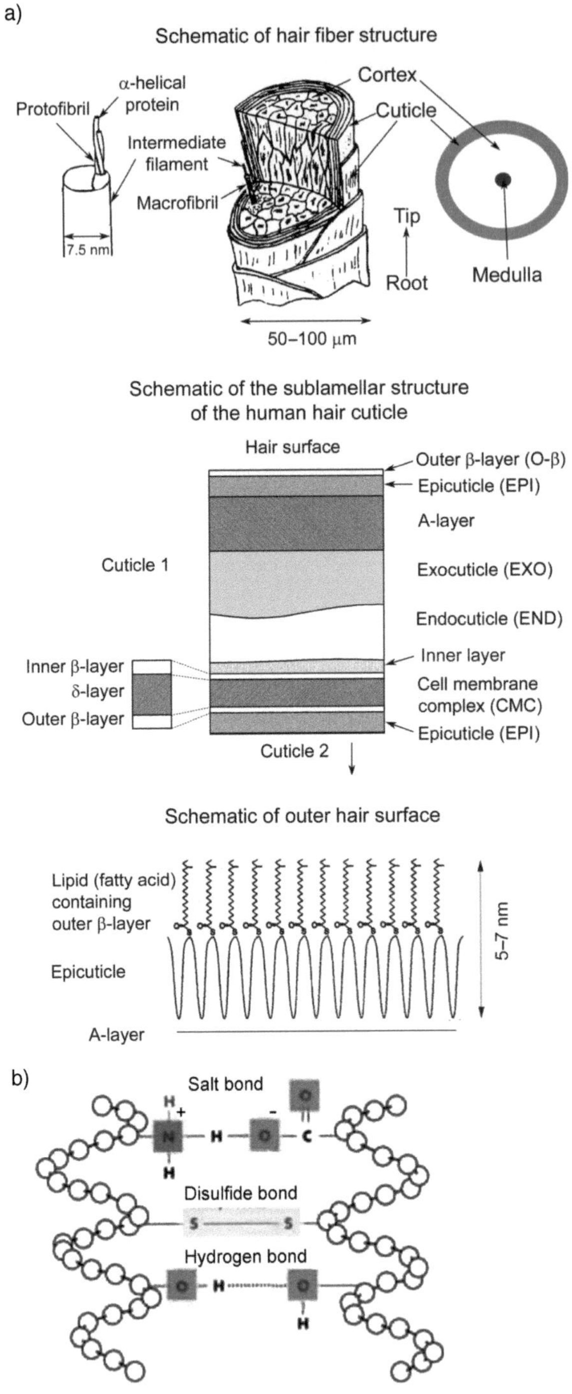

Fig. 1.1 (**a**) Schematic of hair fiber structure and cuticle sublamellar structure (Robbins, 1994; Smith and Swift, 2002) and (**b**) various bonds within hair cellular structure (Bhushan and Chen, 2006; Gray, 2003)

Table 1.1 Summary of chemical species present in human hair

Keratin protein	65–95%
(Amino acids)	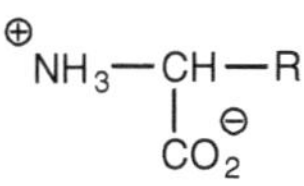 (R: functional group)
Cystine	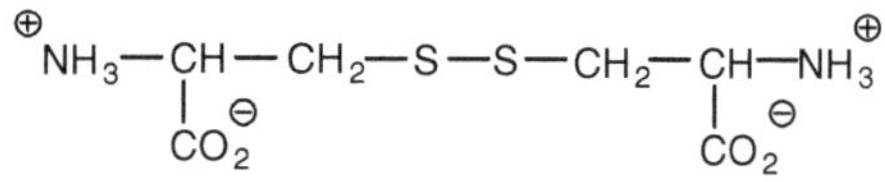
Lipids	Structural and free
18-Methyl eicosanoic acid (18-MEA)	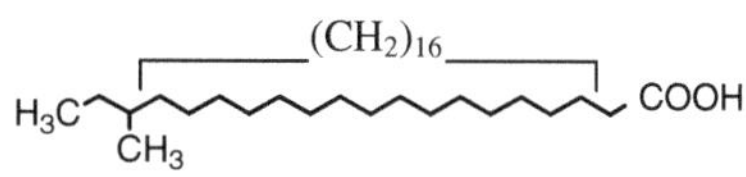
Water	Up to 30%
Pigment and trace elements	Melanin

fibers in a lifetime. Each fiber grows for several years until it falls out and is replaced by a new fiber. Hair typically grows at a rate on the order of 10 mm/month.

1.1.1 The Cuticle

The cuticle consists of flat overlapping cells (scales). The cuticle cells are attached at the root end, and they point toward the tip end of the hair fiber, like tiles on a roof. Each cuticle cell is approximately 0.3–0.5 μm thick, and the visible length of each cuticle cell is approximately 5–10 μm. The cuticle in human hair is generally 5–10 scales thick. Each cuticle cell consists of various sublamellar layers (the epicuticle, the A-layer, the exocuticle, the endocuticle, and the inner layer) and the cell membrane complex (see Fig. 1.1a). Table 1.2 displays the various layers of the cuticle, their respective cystine levels (Robbins, 1994), and other details. The outer epicuticle layer is covered with a thin layer of covalently attached lipid (fatty acid), predominantly 18-methyl eicosanoic acid (18-MEA) (see Table 1.1). This layer constitutes the outer β-layer of the cuticular cell membrane complex, which acts as a boundary lubricant, responsible for low friction and provides a hydrophobic surface. The A-layer is a component of high cystine content (~30%) and located on the outer-facing aspect of each cell. The A-layer is highly cross-linked which gives this layer considerable mechanical toughness and chemical resilience, and the swelling in water is presumed to be minimal. The exocuticle, which is immediately adjacent to the A-layer, is also of high cystine content (~15%). On the inner-facing aspect of each cuticle cell is a thin layer of material which is known as the inner layer. Between the exocuticle and the inner layer is the endocuticle which is low in cystine (~3%). The cell membrane complex (CMC) itself is a lamellar structure,

Table 1.2 Various layers of the cuticle and their details

Cuticle layer	Cystine component	Details
Epicuticle	~12%	18-MEA lipid layer attached to outer epicuticle contributes to lubricity of the hair
A-layer	~30%	Highly cross-linked
Exocuticle	~15%	Mechanically tough Chemically resilient
Endocuticle	~3%	
Inner layer	–	
Cell membrane complex (CMC)	~2%	Lamellar structure Consists of inner β-layer, δ-layer, and outer β-layer

which consists of the inner β-layer, the δ-layer, and the outer β-layer. The outer β-layer of the CMC separates the cuticle cells from each other. Low cohesive forces are expected between the lipid-containing outer β-layer and the δ-layer of CMC, which provides a weak bond. It may result in cuticular delamination during mechanical wear, with the potential advantage of revealing a fresh layer of 18-MEA to the newly exposed surface (Smith and Swift, 2002).

Figure 1.2 shows the SEM images of virgin Caucasian, Asian, and African hair (Wei et al., 2005). It can be seen that Asian hair is the thickest (about 100 μm), followed by African hair (about 80 μm) and Caucasian hair (about 50 μm). The visible cuticle cell is about 5–10 μm long for the three hairs. A listing of various cross-sectional dimensional properties is presented in Table 1.3 (Wei et al., 2005). While Caucasian and Asian hair typically have a similar cross-sectional shape (Asian hair being the most cylindrical), African hair has a highly elliptical shape. African hair is much more curly and wavy along the hair fiber axis than Caucasian or Asian hair.

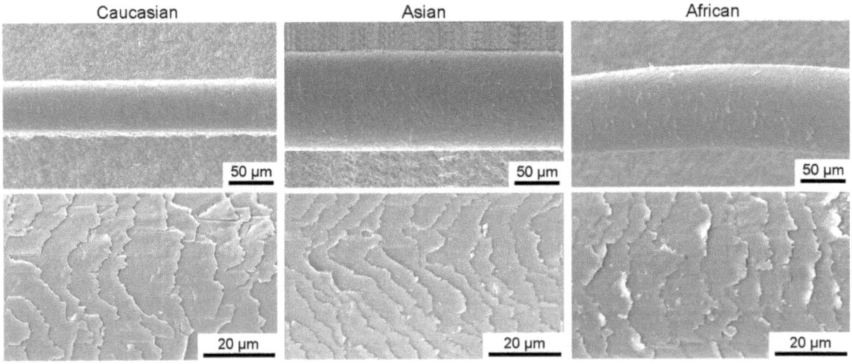

Fig. 1.2 SEM images of various hair (Wei et al., 2005)

Table 1.3 Variation in cross-sectional dimensions of human hair

	Shape	Maximum diameter (D_1) (μm)	Minimum diameter (D_2) (μm)	Ratio D_1/D_2	Number of cuticle scales	Cuticle scale thickness (μm)
Caucasian	Nearly oval	74	47	1.6	6–7	0.3–0.5
Asian	Nearly round	92	71	1.3	5–6	0.3–0.5
African	Oval-flat	89	44	2.0	6–7	0.3–0.5

Average length of visible cuticle scale: about 5–10 μm

Chemically, all ethnic hair is found to have similar protein structure and composition (Dekoi and Jedoi, 1988, 1990; Menkart et al., 1984; Nappe and Kermici, 1989). African hair has less moisture content than Caucasian hair. The shape (diameter, ellipticity, and curliness) of various ethnic hair depends on several factors, including the shape of the hair follicle and its opening; these vary from one person to another and also between races (Gray, 2003; Thibaut et al., 2005). The pronounced ellipsoidal cross section of the hair shaft in African hair could be caused by a heterogeneous and asymmetric fiber framework, in addition to internal mechanical stresses (Thibaut et al., 2005). Previously, it was thought that the elliptical cross section of hair is responsible for curl. While straight hair has circular cross sections (Asian and Caucasian), curly hair has a predominantly elliptical cross section (African). However, recent studies suggest that hair follicle shape and not the cross section is responsible for hair curl (Thibaut et al., 2005). This means that if the follicle is straight, even an elliptical cross section could give rise to straight hair. Both in vitro growth studies and computer-aided 3D reconstruction (Lindelof et al., 1988) support this claim. Curvature of the curly hair is programmed from the basal area of the follicle. This bending process is apparently linked to a lack of symmetry in the lower part of the bulb, affecting the hair shaft cuticle.

Figure 1.3 shows the SEM images of virgin Caucasian hair at three locations: near scalp, middle, and near tip. Three magnifications were used to show the significant differences. The hair near the scalp had complete cuticles, while no cuticles were found on the hair near the tip. This may be because the hair near the tip experienced more mechanical damage during its life than the hair near the scalp. The hair in the middle experienced intermediate damage, i.e., one or more scales of the cuticles were worn away, but many cuticles stayed complete. If some substructures of one cuticle scale, like A-layer or A-layer and exocuticle (see Fig. 1.1a), are gone, or even worse, one or several cuticle scales are worn away, it is impossible to heal the hair biologically, because hair fibers are composed of dead cells. However, it is possible to physically "repair" the damaged hair by using conditioner, one of whose functions is to cover or fill the damaged area of the cuticles. Figure 1.4 shows the high-magnification SEM images of virgin and treated Caucasian hair. The endocuticles (indicated by arrows) were found in both hair. In order for the conditioner to physically repair the hair, it is expected for it to cover the endocuticles. In the case of severely damaged hair, for example, an edge of one whole cuticle scale worn

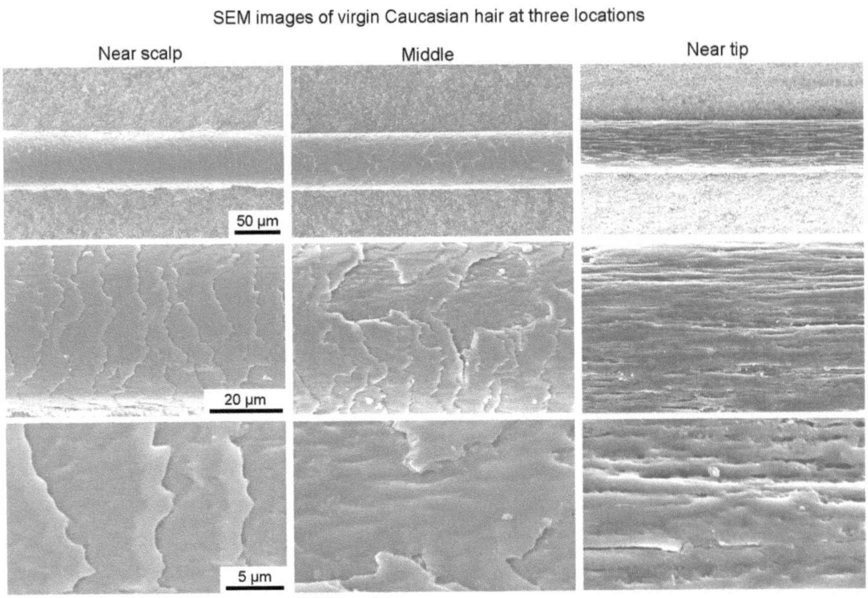

Fig. 1.3 SEM images of virgin Caucasian hair at three locations (Wei et al., 2005)

away, the conditioner may fill that damaged edge. In the SEM image of the treated hair in Fig. 1.4, the substance which stayed near the cuticle edge is probably the conditioner (indicated by an arrow).

Figure 1.5 shows the AFM images of various virgin hair, along with the section plots (LaTorre and Bhushan, 2005a). The arrows point to the position where the section plots were taken from. Each cuticle cell is nearly parallel to the underlying cuticle cell, and they all have similar angles to the hair axis, forming a tile-like hair surface structure. The visible cuticle cell is approximately 0.3–0.5 μm thick and about 5–10 μm long for all three hairs.

1.1.2 The Cortex and Medulla

The cortex contains cortical cells and the intercellular binding material, or the cell membrane complex. The cortical cells are generally 1–6 μm thick and 100 μm long, which run longitudinally along the hair fiber axis and take up the majority of the inner hair fiber composition (Randebrook, 1964). The macrofibrils (about 0.1–0.4 μm in diameter) comprise a major portion of the cortical cells. Each macrofibril consists of intermediate filaments (about 7.5 nm in diameter), previously called microfibrils, and the matrix. The intermediate filaments are low in cystine (~6%), and the matrix is rich in cystine (~21%). The cell membrane complex consists of cell membranes and adhesive material that binds the cuticle and cortical cells together. The intercellular cement of the cell membrane complex is primarily

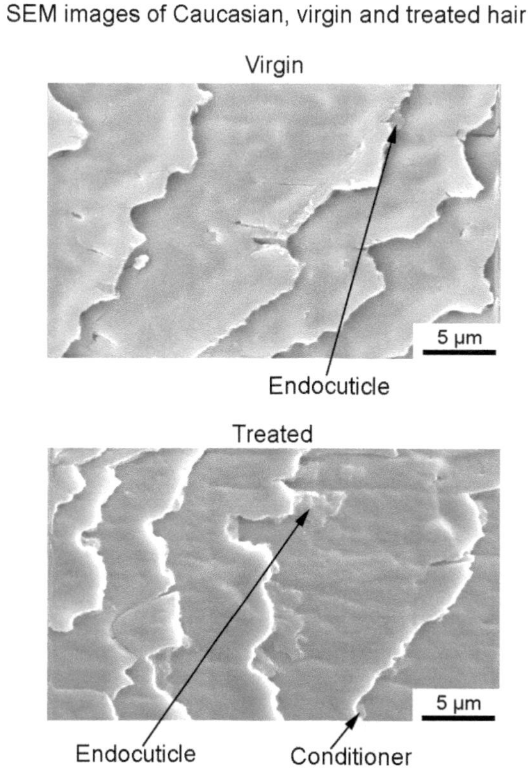

Fig. 1.4 SEM images of Caucasian, virgin and treated hair (Wei et al., 2005)

non-keratinous protein and is low in cystine content (~2%). The medulla of human hair, if present, generally makes up only a small percentage of the mass of the whole hair and is believed to contribute negligibly to the mechanical properties of human hair fibers.

Figure 1.6a shows the SEM images of virgin hair cross section (Wei et al., 2005) and Fig. 1.6b shows the TEM images of a cross section of human hair (Swift, 1997).

1.2 Skin

Skin covers and protects our bodies. The skin at the forehead and scalp areas are of most interest when dealing with human hair, since most hair care products are developed specifically for head hair. The skin of the hand and fingers is also of importance because the "feel" of hair is often sensed by physically touching the fibers with these regions. In general, skin is composed of three main parts: epidermis, dermis, and subcutaneous tissue; see Fig. 1.7. The subcutis lies under the dermis and consists

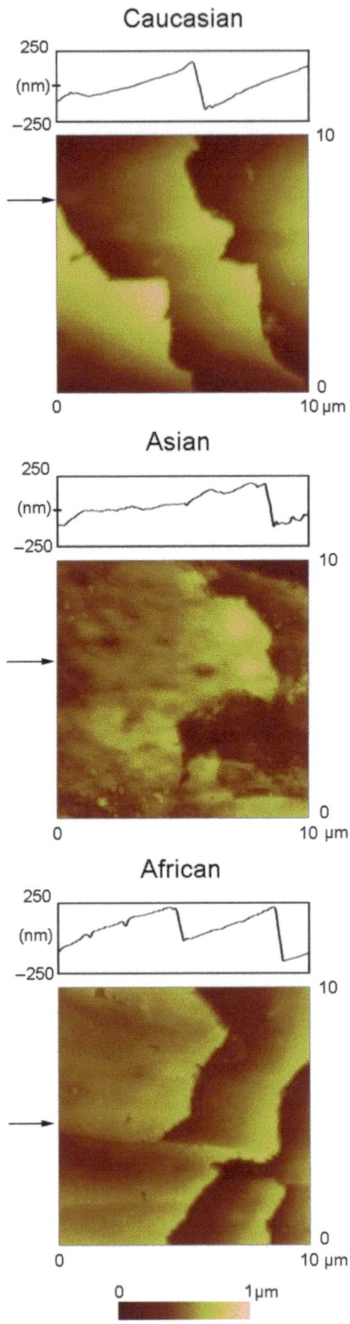

Fig. 1.5 AFM images of various virgin hair (LaTorre and Bhushan, 2005a)

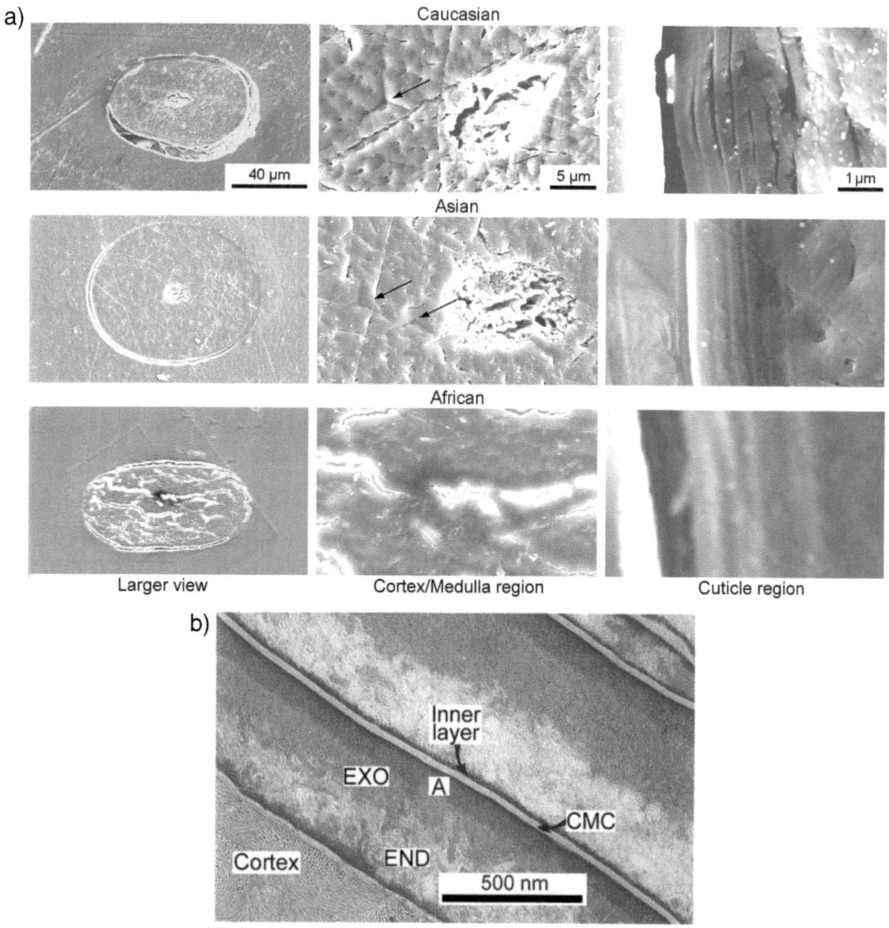

Fig. 1.6 (**a**) SEM images of virgin hair cross section (Wei et al., 2005); (**b**) TEM of hair cross section (in the figure EXO, END, and CMC stand for exocuticle, endocuticle, and cell membrane complex, respectively) (Swift, 1997)

of adipose tissue or fat cells with collagen partition. The dermis lies below the epidermis and supports it structurally and nutritionally. It contains blood vessels, nerves, hair follicles, arrector pili muscle, sweat glands, and sebaceous glands. The epidermis is the outer layer of the skin. It contains four distinct cellular layers: basal layer, spinous layer, granular layer, and keratin layer. Cells proliferate in the basal layer of the epidermis. Upon leaving the basal layer, cells start to differentiate and migrate upward through the spinous layer and granular layer, finally reaching the keratin layer, from which they are shed. The keratin layer is about 10–20 μm thick and is composed of 10–15 layers of anuclear, keratin-rich corneocytes imbedded in an extracellular lipid matrix (Lodén and Maibach, 2000; Shai et al., 2001). It is the keratin layer which comes in contact with cosmetic products, fabric, and other surfaces. It serves as a penetration, dehydration, and protection barrier against various environmental hazards. Water is a crucial factor to keratin layer barrier function and structure (Leyden and Rawlings, 2002). In soft and flexible skin, the water content of the keratin layer is between 10 and 30% (Elsner et al., 1994). If insufficient water remains in the keratin layer, it leads to epidermal hyperplasia, mast cell degranulation, and cytokine secretion. These issues are considered to be harmful to the requirement for healthy and desirable skin.

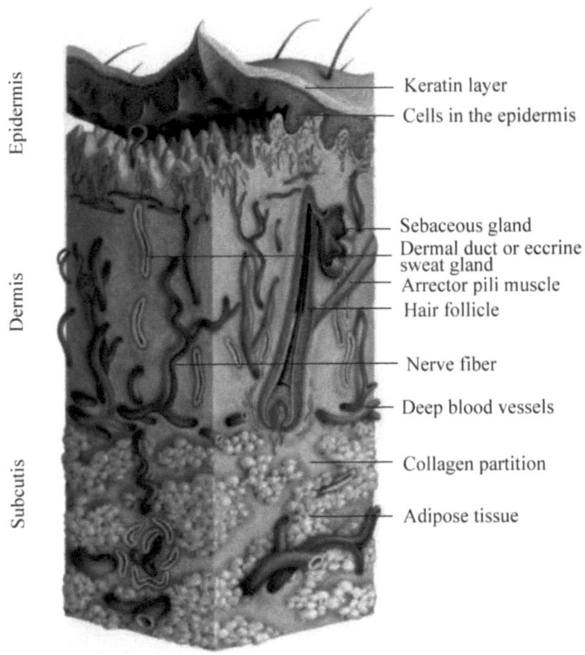

Fig. 1.7 Schematic image of human skin structure with different layers: subcutis, dermis, and epidermis (Shai et al., 2001)

1.3 Hair Care: Cleaning and Conditioning Treatments, and Damaging Processes

Everybody wants beautiful, healthy hair and skin. For most people, grooming and maintenance of hair and skin is a daily process. The demand for products that improve the look and feel of these surfaces has created a huge industry. Beauty care technology has advanced the cleaning, protection, and restoration of desirable hair properties by altering the hair surface. For many years, especially in the second half of the twentieth century, scientists have focused on the physical and chemical properties of hair to consistently develop products which alter the health, feel, shine, color, softness, and overall aesthetics of hair. Hair care products such as shampoos and conditioners aid the maintenance and grooming process. Shampoos clean the hair and skin oils, and conditioners repair hair damage and make the hair easier to comb; prevent flyaway; and add feel, shine, and softness. Mechanical processes such as combing, cutting, and blowdrying serve to style the hair. Chemical products and processes such as chemical dyes, colorants, bleaches, and permanent wave treatments enhance the appearance and hue of the hair. Of particular interest is how all these common hair care items deposit onto hair and change its properties, since these properties are closely tied to product performance. The fact that companies like Procter & Gamble, L'Oreal, and Unilever have hair care product sales consistently measured in billions of US dollars (http://www.pg.com; http://www.loreal.com; http://unilever.com) suggests that understanding the science behind human hair has more than just purely academic benefits, as well.

While products and processes such as combing, chemical dyeing, and permanent wave treatment are used to enhance appearance and style of the hair, they also contribute a large amount of chemical and mechanical damage to the fibers, which leads to the degradation of structure and mechanical properties. As a result, the fibers become weak and more susceptible to breakage after time, which is undesirable for healthy hair. Shampoos and conditioners which typically serve cleaning and repairing functions to the hair surface, respectively, have a distinct effect on mechanical properties as well.

The tribology of the hair also changes as a function of the various hair care products and processes. Figure 1.8 illustrates schematically various functions, along with the macro- and micro/nanoscale mechanisms behind these interactions that make surface roughness, friction, and adhesion very important to hair and skin (LaTorre and Bhushan, 2005a). Desired features and corresponding tribological attributes of conditioners are listed in Table 1.4 (LaTorre et al., 2006). For a smooth, wet and dry feel, friction between hair and skin should be minimized in wet and dry environments, respectively. For a good feel with respect to bouncing and shaking of the hair during walking or running, friction between hair fibers and groups of hair fibers should be low. The friction one feels during combing is a result of interactions between hair and the comb material (generally a plastic), and this too needs to be low to easily maintain, sculpt, and comb the hair. To minimize entanglement, adhesive force (the force required to separate the hair fibers) needs to be low. In other cases, a certain level of adhesion may be acceptable and is often a function of the hair style. For individuals seeking "hair alignment," where hair fibers lay flat and parallel to each other, a small amount of adhesive force between fibers may be desired. For more complex and curly styles, even higher adhesion between fibers may be optimal.

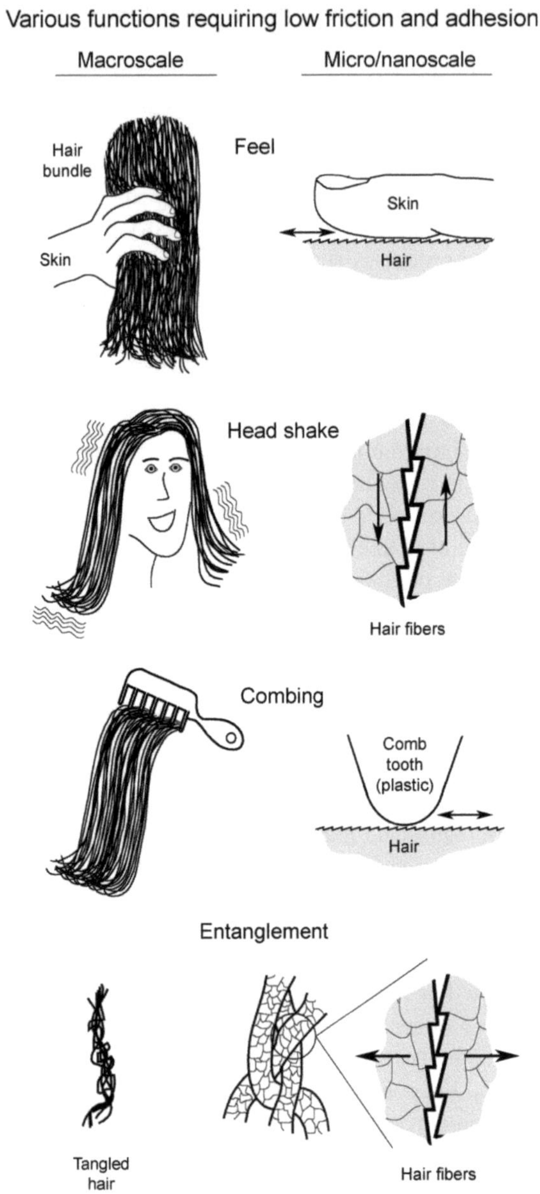

Fig. 1.8 Schematics illustrating various functions with associated macroscale and micro/nanoscale mechanisms of hair and skin friction during feel or touch, shaking and bouncing of the hair, combing, and entanglement (LaTorre and Bhushan, 2005a)

Table 1.4 Desired features and corresponding tribological attributes of conditioners

Desired hair feature	Tribological attributes
Smooth feel in wet and dry environments	Low friction between hair and skin in respective environment
Shaking and bouncing during daily activities	Low friction between hair fibers and groups of hair
Easy combing and styling	Low friction between hair and comb (plastic) and low adhesion. Note: More complex styles may require higher adhesion between fibers

1.3.1 Cleaning and Conditioning Treatments: Shampoo and Conditioner

Shampoos are used primarily to clean the hair and scalp of dirt and other greasy residue that can build up after time. Shampoos also have many secondary functions including controlling dandruff, reducing irritation, and even conditioning. Conditioners, on the other hand, are used primarily to give the hair a soft, smooth feel which results in easier hair combing. Secondary functions include preventing "fly-away" hair due to static electricity, giving the hair a shiny appearance, and protecting the hair from further damage by forming a thin coating over the fibers.

Further developments in marketing and aesthetic factors (brand name, fragrance, feel, and color of the shampoos and conditioners) have created new market segments. In many instances, these factors have become primary reasons for use.

1.3.1.1 Shampoo: Constitution and Main Functions

The following discussion is based on Gray (2001, 2003). As stated above, shampoos serve various cleaning functions for the hair and scalp. In the past, typical shampoos were mainly soap-based products. However, soaps did not have very good lathering capability and often left a residual "scum" layer on the hair that was undesirable and could not be rinsed off.

In modern shampoos, advances in chemistry and technology have made it possible to replace the soap bases with complex formulas of cleansing agents, conditioning agents, functional additives, preservatives, aesthetic additives, and even medically active ingredients. Clarifying shampoo (with no conditioner), typically used for full volume with lot of air between hair, is transparent. The pH value is typically 5–6. Table 1.5 shows the most common ingredients of shampoos and their functions.

Cleansing Agents: In most modern shampoos, the primary cleansing agents are anionic surfactants which are known to be good in cleaning. Dirt and greasy residue are removed from the hair and scalp by these surfactants, making them the most important part of the shampoo. Surfactants have great lathering capabilities and rinse off very easily; see Table 1.5 for a full list of features.

Table 1.5 Components of common shampoos and their functions

Shampoo component	Functions
Cleansing agents	– Produce lather to trap greasy matter and prevent re-deposition – Remove dirt and grease from hair and scalp – Stabilize the mixture and help keep the ingredient network together – Thicken the shampoo to the desired viscosity
Conditioning agents	Condition the hair
Functional additives	Control the viscosity and pH levels of the shampoo
Preservatives	Prevent decomposition and contamination of shampoo
Aesthetic additives	Enhance color, scent, and luminescence of shampoo
Medically active ingredients	Aid treatment of dandruff or hair loss

Surfactant molecules have two different ends, one which is negatively charged and soluble in water (unable to mix with greasy matter) and another which is soluble in greasy matter (unable to mix with water). In general, surfactants clean the hair by the following process: Surfactant molecules encircle the greasy matter on the hair surface. The molecule end which is soluble in greasy matter buries itself in the grease, which leaves the water-soluble molecule end to face outward with a negative charge. Since the hair fibers are negatively charged as well, the two negative charges repel each other. Thus, the greasy matter is easily removed from the hair surface and rinsed off.

Shampoos contain several surfactants, generally up to four, which clean the hair differently depending on the hair type of the individual. Mild cleansing systems, which do not damage or irritate the scalp, hair, and eyes, are now quite common.

Conditioning Agents: Many shampoos contain conditioning agents which serve many of the same roles as full conditioners. Conditioning agents are further described in the following subsection.

Functional Additives: Functional additives can aid in controlling the thickness and feel of the shampoo itself. Simply stated, the right blend is required so that the shampoo is not too thin and not too thick. Functional additives can also control the acidity of the shampoo by obtaining a goal pH level, typically around a value of 4.

Preservatives: Preservatives resist germs and prevent decomposition of the shampoos. They also prevent various other health risks that accompany contamination by germs and bacteria. Typical preservatives in shampoos are sodium benzoate, parabens, 1,3-dimethylol-5,5-dimethyl (DMDM) hydantoin, and tetrasodium ethylenediamine tetraacetic acid (EDTA).

Aesthetic Additives: Shampoos contain many aesthetic additives which enhance the appearance, color, and smell of the mixture. These additives typically give the shampoo the luminous shine and pleasant fragrance to which many consumers are accustomed.

Medically Active Ingredients: For people with dandruff and other more serious hair and scalp disorders, shampoos are available with active ingredients which aim to treat or control these conditions. In the treatment of dandruff, zinc pyrithione is a common shampoo additive. For hair loss issues, panthenol is commonly added to shampoos to aid in hair growth and moisture content.

1.3.1.2 Conditioner: Constitution and Main Functions

As stated earlier, many shampoos have certain levels of conditioning agents which mimic the functions of a full conditioner product. Conditioner molecules contain cationic surfactants which give a positive electrical charge to the conditioner. The negative charge of the hair is attracted to the positively charged conditioner molecules, which results in conditioner deposition on the hair, see Fig. 1.9. This is especially true for damaged hair, since damaging processes result in hair fibers being even more negatively charged. The attraction of the conditioner to hair results in a reduction of static electricity on the fiber surfaces, and consequently a reduction in the "flyaway" behavior. The conditioner layer also flattens the cuticle scales against each other, which improves shine and color. The smooth feel resulting from conditioner use gives easier combing and detangling in both wet and dry conditions (see Table 1.4).

Conditioner consists of a gel network chassis (cationic surfactants, fatty alcohols, and water) for superior wet feel and a combination of conditioning actives (cationic surfactants, fatty alcohols, and silicones) for superior dry feel. Figure 1.10 shows the transformation of the cationic surfactants and fatty alcohol mixture into the resulting gel network, which is a frozen lamellar liquid crystal gel phase (Bhushan and Chen, 2006). The process starts as an emulsion of the surfactants and alcohols in water. The materials then go through a strictly controlled heating and cooling process: the application of heat causes the solid compounds to melt, and the solidification process enables a setting of the lamellar assembly molecules in a fully extended conformation, creating a lamellar gel network. When this network interacts with the

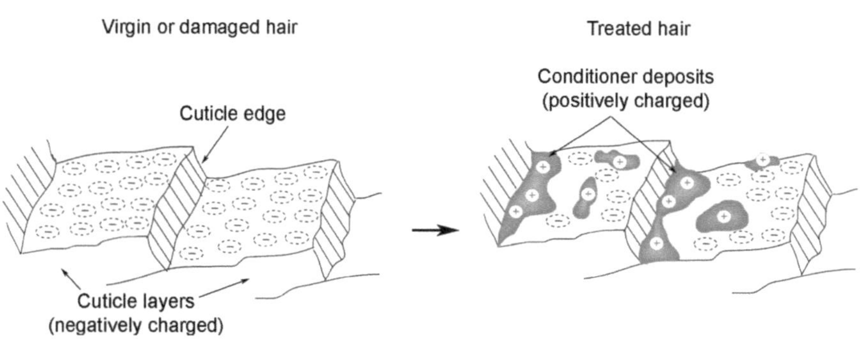

Fig. 1.9 Negatively charged hair and the deposition of positively charged conditioner on the cuticle surface (LaTorre et al., 2006)

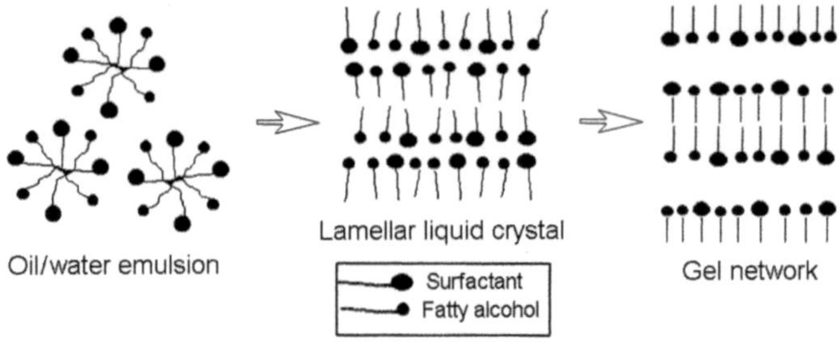

Fig. 1.10 Conditioner formation from emulsion to gel network (Bhushan and Chen, 2006)

hair surface, the high concentration of fatty alcohols make it the most deposited ingredient group, followed by the silicones and cationic surfactants. Typical deposition levels for cationic surfactant, fatty alcohol, and silicone are around 500–800, 1000–2000, and 200 ppm, respectively. Typical concentrations are approximately 2–5, 5–10, and 1–10 wt%, respectively (LaTorre et al., 2006).

The benefits of the conditioner are shown in Table 1.6 (LaTorre et al., 2006). The wet feel benefits are creamy texture, ease of spreading, slippery feel while applying, and soft rinsing feel. The dry feel benefits are moistness, softness, and easier dry combing. Each of the primary conditioner ingredients also has specific functions and roles that affect performance of the entire product. Table 1.7 displays the functions of the major conditioner ingredients and their chemical structure (LaTorre et al., 2006). Cationic surfactants are critical to the forming of the lamellar gel network in conditioner and also act as a lubricant and static control agent, since their positive charge aids in counteracting the negative charge of the hair fibers. Fatty alcohols are used to lubricate and moisturize the hair surface, along with forming the gel network. Finally, silicones are the main source of lubrication in the conditioner

Table 1.6 Combinations of conditioner ingredients and their benefits toward wet and dry feel

Gel network chassis for superior wet feel	
Key Ingredients	Benefits
– Cationic surfactant – Fatty alcohols – Water	– Creamy texture – Ease of spreading – Slippery applying feel – Soft rinsing feel
Combination of "conditioning actives" for superior dry feel	
Key Ingredients	Benefits
– Silicones – Fatty alcohols – Cationic surfactant	– Moist – Soft – Dry combing ease

Table 1.7 Chemical structure and purpose/function of conditioner ingredients

Ingredient		Chemical structure	Purpose/function
Water			
Cationic surfactants	Stearamidopropyl dimethylamine	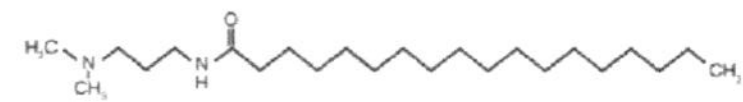	– Aids formation of lamellar gel network – Lubricates and is a static control agent
	Behenyl amidopropyl dimethylamine glutamate (BAPDMA)	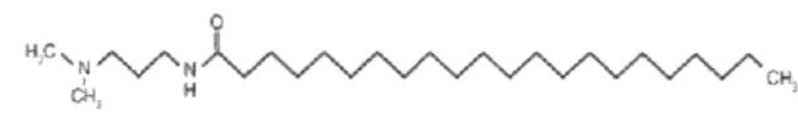	
	Behentrimonium chloride (BTMAC) $CH_3(CH_2)_{21}N(Cl)(CH_3)_3$	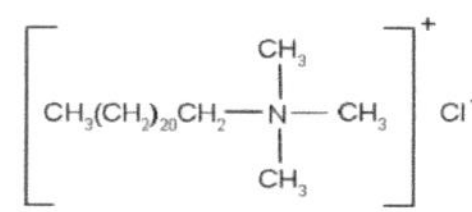	
Fatty alcohols	Stearyl alcohol ($C_{18}H_{37}OH$)		– Lubricates and moisturizes – Aids formation of lamellar gel network along with cationic surfactant
	Cetyl alcohol ($C_{16}H_{33}OH$)		
Silicones	Polydimethylsiloxane (PDMS) blend (Dimethicone)	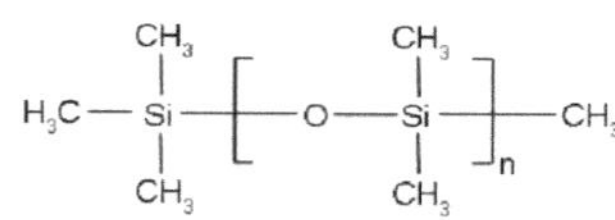	– Primary source of lubrication – Gives hair a soft and smooth feel

formulation. Conditioners are opaque because of silicone particles. The pH value is low, about 3. Compositions are ethnic based. For example, Asian hair have large diameter; therefore, a higher concentration of silicon is used as compared to that for Caucasian hair (3–5% vs. 0.5%).

1.3.2 Damaging Processes

In Sect. 1.3.1 we discussed some of the products which aid in "treating" the hair. There are other hair care products and processes which, while creating a desired look or style to the hair, also bring about significant damage to the fibers. Most of these processes occur on some type of periodic schedule, whether it be daily (while combing the hair) or monthly (haircut and coloring at a salon). In general, hair fiber damage occurs most readily by mechanical or chemical means or by a combination of both (chemo-mechanical).

1.3.2.1 Mechanical Damage

Mechanical damage occurs on a daily basis for many individuals. The damage results from large physical forces or temperatures which degrade and wear the outer cuticle layers. Common causes are

- combing (generally with plastic objects, and often multiple times over the same area lead to scratching and wearing of the cuticle layers)
- scratching (usually with fingernails around the scalp)
- cutting (affects the areas surrounding the fiber tips)
- blowdrying (high temperatures thermally degrade the surface of the hair fibers)

1.3.2.2 Permanent Wave Treatment

Permanent wave treatments saw many advances in the beginning of the twentieth century, but have not changed much with the invention of the Cold Wave around the turn of that century. Generally speaking, the Cold Wave uses mercaptans (typically thioglycolic acid) to break down disulfide bridges and style the hair without much user interaction (at least in the period soon after the perm application) (Gray, 2001). The Cold Wave process does not need increased temperatures (so no thermal damage to the hair), but generally consists of a reduction period (whereby molecular reorientation to the cuticle and cortex occurs via a disulfide–mercaptan interchange pathway; Robbins, 1994) followed by rinsing, setting of the hair to the desired style, and finally neutralization to decrease the mercaptan levels and stabilize the style. The chemical damage brought on by the permanent wave can increase dramatically when not performed with care.

1.3.2.3 Chemical Relaxation

Commonly used as a means of straightening hair (especially in highly curved, tightly curled African hair), this procedure uses an alkaline agent, an oil phase, and a water phase of a high-viscosity emulsion to relax and reform bonds in extremely curly hair. A large part of the ability to sculpt the hair to a desired straightness comes from the breakage of disulfide bonds of the fibers.

1.3.2.4 Coloring and Dyeing

Hair coloring and dyeing have become extremely successful hair care procedures, due in part to "over-the-counter" style kits which allow home hair care without professional assistance. The most common dyes are para dyes, which contain para-phenylenediamine (PPD) solutions accompanied by conditioners and antioxidants. Hydrogen peroxide (H_2O_2) is combined with the para dyes to effectively create tinted, insoluble molecules which are contained within the cortex and are not small enough to pass through the cuticle layers, leaving a desired color to the hair. Due to the levels of hydrogen peroxide, severe chemical damage can ensue in the cuticle and cortex.

Bleaching: Like dyeing, bleaching consists of using hydrogen peroxide to tint the hair. However, bleaching can only lighten the shade of hair color, as the H_2O_2 releases oxygen to bind hair pigments (Gray, 2001). Bleaching may also be applied to limited areas of the hair (such as in highlights) to create a desired look. The chemical damage brought on by bleaching leads to high porosity and severe wear of the cuticle layer.

1.4 Organization of the Book

In this book, we present a comprehensive study of various hair and skin structural, nanomechanical, and nanotribological properties as a function of ethnicity, damage, conditioning treatment, and various environments. Various cellular structures (such as the cortex and the cuticle) of human hair and fine sublamellar structures of the cuticle, such as the A-layer, the exocuticle, the endocuticle, and the cell membrane complex, are easily identified and studied. Nanomechanical properties including hardness, elastic modulus, tensile deformation, fatigue, creep, and scratch resistance are discussed. Nanotribological properties including surface roughness, friction, adhesion, and wear are presented, as well as investigations of conditioner localization and thickness. To understand the electrostatic charge buildup on hair, surface potential studies are presented.

Chapter 2
Experimental Methods

Early research into human hair was done primarily on the chemical and physical properties of the hair fiber itself. Key topics dealt with the analysis of chemical composition in the fiber, microstructure, and hair growth, to name a few. Until about 2000, most information about the detailed structure of human hair was obtained from scanning electron microscope (SEM) and transmission electron microscope (TEM) observations (Robbins, 1994; Swift, 1991, 1997; Wei et al., 2005). Mechanical properties were also of interest. Most of the mechanical property measurements of human hair were on the macroscale and used conventional methods, such as tension, torsion, and bending tests (Barnes and Roberts, 2000; Feughelman, 1997; Robbins, 1994; Swift, 1999, 2000; Jachowitz and McMullen, 2002). The mechanical properties obtained from these tests are the overall mechanical properties of the hair, not just of the hair surface. Efforts were also made to study the effects of environmental and chemical damage and treatment, such as dyeing, bleaching, and polymer application; these topics have remained a mainstream area of investigation due to the availability and formation of new chemicals and conditioning ingredients. Tribology of human hair has generally been studied via macroscale friction force of hair. As a matter of fact, much of the tribological work performed by the hair care industry today still focuses on the measurement of macroscale friction, particularly between a skin replica and a hair swatch of interest (Robbins, 1994). The intrinsic differences of the hair as a function of ethnicity eventually became a concern as well. For instance, research has shown that African-American hair has higher resistance to combing, higher static charge, and lower moisture content than Caucasian hair (Syed et al., 1995). Because of differences like these, a growing number of hair care products specifically tailored for ethnic hair care have been developed and sold with commercial success.

Modern research since the late 1990s has been primarily concerned with using micro/nanoscale experimental methods such as atomic force/friction force microscopies (AFM/FFM) and nanoindentation to answer the complex questions surrounding the structure and behavior of hair (Bhushan, 2008a). Nanoscale characterization of the cellular structure, mechanical properties, and tribological properties of hair are essential to evaluate and develop better cosmetic products and

B. Bhushan, *Biophysics of Human Hair*, Biological and Medical Physics, Biomedical Engineering, DOI 10.1007/978-3-642-15901-5_2,

to advance the understanding of composite biological systems, cosmetic science, and dermatology. AFM/FFM have been used to effectively study the structure of hair surface and cross section. AFM provides the potential for being able to image the cellular structure and molecular assembly of hair, for determining various properties of hair, such as elastic modulus and viscoelastic properties, and for investigating the physical behavior of various cellular structures of hair in various environments (Bhushan and Chen, 2006; Chen and Bhushan, 2005). As a non-invasive technique, AFM has been used to evaluate the effect of hair treatment and can be operated in ambient conditions in order to study the effect of environment on various physical properties. AFM studies on hair fibers have been carried out for surface topographic imaging (LaTorre and Bhushan, 2005a; Smith and Swift, 2002) and friction, adhesion, and wear properties of hair and skin and the effects of hair care products on hair (Chen and Bhushan, 2006; LaTorre and Bhushan, 2005a, b, 2006; LaTorre et al., 2006; Lodge and Bhushan, 2006a). Roughness parameters have been measured to compare changes due to damaging processes. Friction force has been measured to understand damage or conditioner distribution and its effect on hair tribology. Adhesive force mapping has been useful to observe the conditioner distribution as well. Surface charge of human hair has a significant effect on manageability, flyaway behavior, feel, and appearance. It is known that interaction of hair with dissimilar materials, such as plastic combs, hands, and latex balloons, creates a charge on hair (Lodge and Bhushan, 2007a, b). Physical wear has been shown to cause surface potential change in conductors and semiconductors (Bhushan and Goldade, 2000a, b; DeVecchio and Bhushan, 1998), including hair (Lodge and Bhushan, 2007a, b; Seshadri and Bhushan, 2008c).

The nanoindenter has been used to characterize the nanomechanical behavior of the hair surface and cross section using nanoindentation and nanoscratch techniques (Wei and Bhushan, 2006; Wei et al., 2005). These properties are important for evaluating cosmetic products by comparing the nanomechanical properties, such as hardness, elastic modulus, and scratch resistance, of the hair surface before and after chemical damage or conditioner treatment. Since hair is a nanocomposite biological fiber with well-characterized structures, which will be described in detail in the next section, it is a good model to study the role of various structural and chemical components in providing mechanical strength to composite biological fibers (Wei et al., 2005). Furthermore, the quantitative determination of the mechanical properties of human hair can also provide dermatologists with some useful markers for the diagnosis of hair disorders and for the evaluation of their response to therapeutic regimens (Nikiforidis et al., 1992; Swanbeck et al., 1970). Combing results in physical damage such as scratching, and therefore scratch resistance is useful to study, especially when conditioner is applied. Human hair fibers stretch and consequently experience tensile stresses as they are brushed, combed, and go through the styling process. Hence the behavior of hair under tension is of interest. On human hair, characterization of fracture after deformation has been carried out using SEM and light/fluorescence microscopy (Henderson et al., 1978; Swift, 1999). With in situ experiments, it is possible to systematically follow the progress of morphological change and deformation in the material and to accurately pinpoint the initiation of major deformation events. In situ tensile loading experiments in an AFM have been carried out to study the progress of deformation and morphological changes in the hair fiber, by pausing the straining at regular intervals (Seshadri and Bhushan, 2008a, b).

2.1 Experimental Apparatuses

Table 2.1 shows a comparison between the various methods used to study hair on the micro/nanoscale. The SEM has long been the standard means of investigating the surface topography of human hair. The SEM uses an electron beam to give a high-resolution image of the sample, but cannot provide quantitative data regarding the surface. SEM requires the hair sample to be covered with a very thin layer of a conductive material and needs to be operated under vacuum during both metallization and measurements. Surface metallization and vacuum exposure could potentially induce modifications to the surface details. TEM examinations provide fine detailed internal structure of human hair. However, thin sections of 50–100 nm thickness and heavy metal compounds staining treatment are required for TEM examinations. The cutting of these thin sections with the aid of an ultra-microtome is not an easy task. Moreover, since both SEM and TEM techniques cannot be used to measure the physical properties (e.g., mechanical and tribological) of various cellular structures of human hair of interest and do not allow ambient imaging conditions, many outstanding issues remain to be answered. For example: How do the various cellular structures of hair behave physically in various environments (temperature, humidity, etc.)? How do they swell in water? For conditioner-treated hair, how thick is the conditioner layer and how is the conditioner distribution on hair surface?

AFM is now commonly used for morphological, structural, tribological, and mechanical characterization of surfaces (Bhushan, 1999a, b, 2002, 2008b). As a non-invasive technique, AFM has been used to evaluate the effect of hair treatment and can be used in ambient conditions to study the effect of environment on various physical properties. A schematic of an AFM imaging a hair fiber is shown in Fig. 2.1. AFM/FFM uses a sharp tip with a radius of approximately 10–50 nm. This significant reduction in tip to sample interaction compared to the macroscale allows the simulation of single asperity contact to give detailed surface information. Contact mode allows simultaneous measurement of surface roughness and friction force. Different AFM operating modes, tapping mode and torsional resonance (TR) mode, can be used for measurements of material stiffness and viscoelastic properties mapping using amplitude and phase angle imaging. To study the electrostatic charge buildup on hair, surface potential studies can be carried out using AFM as a nano-Kelvin probe.

When skin comes in contact with hair, actual contact occurs over a large number of asperities. During relative motion, friction and adhesion are governed by

Table 2.1 Comparison of methods used to characterize hair on micro/nanoscale

Method	Type of information	Quantitative data	Normal load	Resolution (nm)	Limitations
Scanning electron microscopy (SEM)	Structural	Gross dimensions	None	0.2–1 (spatial)	– Requires thin conductive coating on sample – Requires vacuum environment – Expensive instrumentation[a] – Tedious[a]
Transmission electron microscopy (TEM)	Structural	Gross dimensions	None	0.2–1 (spatial)	– Requires thin sections (< 100 nm) and heavy metal compound staining treatment – Requires vacuum environment – Expensive instrumentation[a] – Tedious[a]
Atomic force/friction force microscopy (AFM/FFM)	Structural, mechanical, and tribological	– Surface roughness – Elastic modulus – Viscoelasticity – Friction – Adhesion – Conditioner thickness	< 0.1–500 nN[a]	0.2–1 (spatial)[a] 0.02 (vertical)[a]	None
Nanoindenter	Mechanical	– Hardness – Elastic modulus – Creep – Scratch resistance	< 0.1–350 mN	400 (spatial)[b] 0.1 (vertical)[b]	None

[a] Bhushan (2002)
[b] Bhushan and Li (2003)

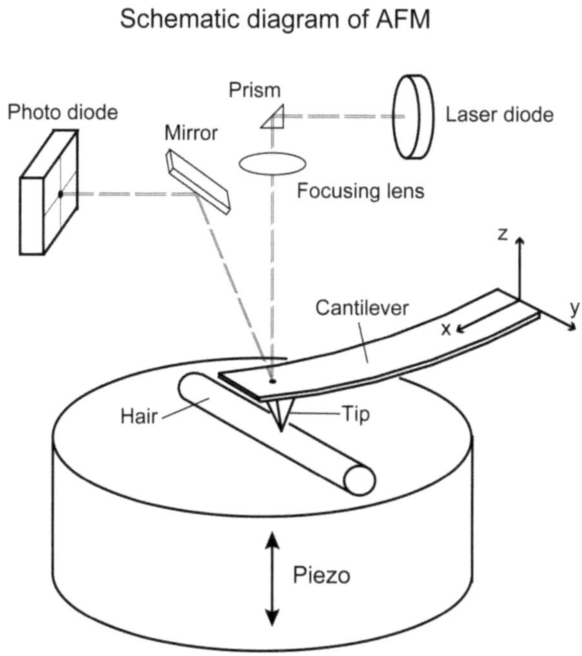

Fig. 2.1 Schematic diagram of AFM operation with human hair sample

the surface interactions which occur at these asperities. Until about 2000, much of the work in the industry had focused on the measurement of macroscale friction, particularly between a skin replica and a hair swatch of interest (Robbins, 1994). Figure 2.2 shows schematics of typical macro- and micro/nanoscale test apparatuses. However, there are many problems associated with performing these types of measurements. Factors such as topographical variations, lumping of the hair fibers, the large size of the synthetic skin, and traditional measurement system errors can all lead to uncertainty in the data.

Depth-sensing nanoindentation measurement techniques are now commonly used to measure nanomechanical properties of surface layers of bulk materials and of ultrathin coatings (Bhushan, 1999b, 2008b; Bhushan and Li, 2003). More recently, nanoindentation technique has been used to investigate nanomechanical properties (such as hardness, Young's modulus, creep, and scratch resistance) of various cellular layers of glass fibers, keratin fibers, and hair surface/cross section (Li et al., 1996; Parbhu et al., 1999; Wei et al., 2005; Wei and Bhushan, 2006). In situ tensile loading experiments in the AFM have been carried out to study the progress of deformation and morphological changes in the hair fiber (Seshadri and Bhushan, 2008a, b).

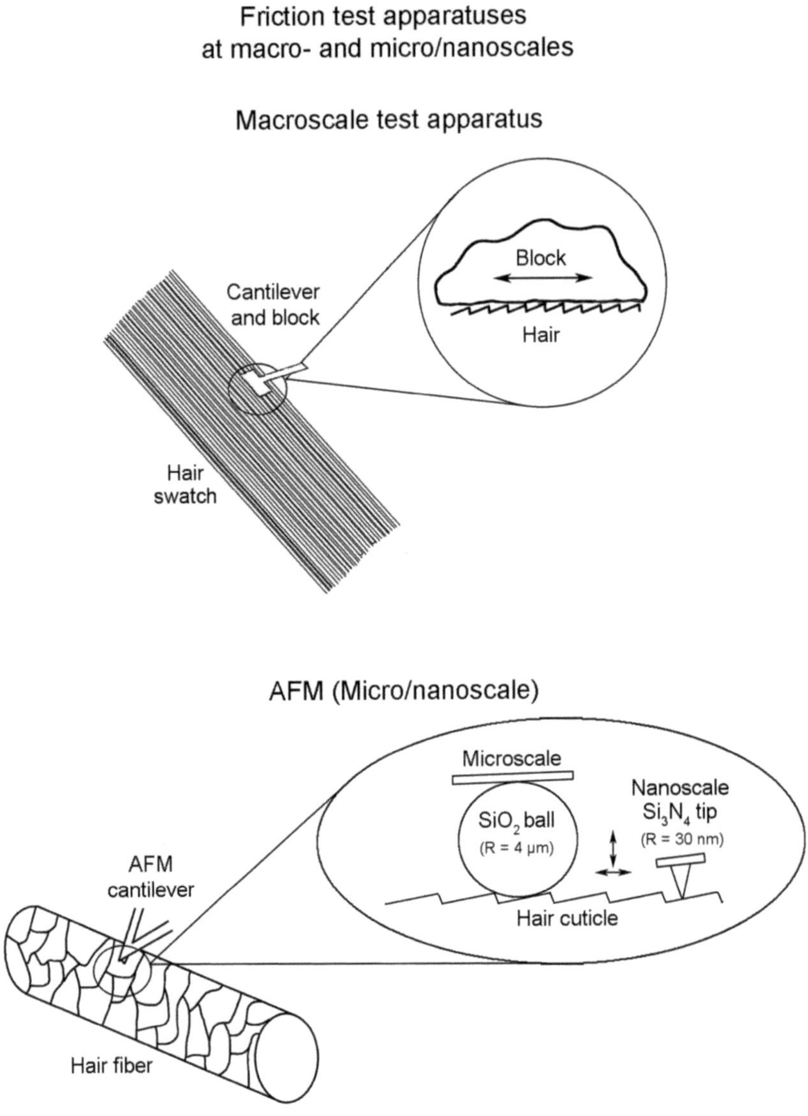

Fig. 2.2 Friction test apparatuses at macro- and micro/nanoscales

2.2 Experimental Procedure

2.2.1 Structural Characterization Using an AFM

An AFM (MultiMode Nanoscope IIIa, DI-Veeco, Santa Barbara, CA) with modifications for the TR mode and a nanoscope Extender Electronic circuit for the measurement of the phase angle were used in the study reported in the book (Bhushan and Chen, 2006; Chen and Bhushan, 2005). All measurements are conducted in ambient conditions (22 ± 1°C, 50 ± 5% relative humidity). The probes used in this study were single beam etched Si probes (MikroMasch) with a fundamental flexural mode frequency of 75 kHz and a fundamental torsional resonance frequency of 835 kHz with a quality factor around 1000. The dimension of the cantilever is typically 230 μm × 40 μm × 3 μm with a flexural spring constant of 1–5 N/m and a torsional spring constant estimated to be 30–150 N/m. The radius of curvature of the tip is about 10 nm. Surface height images shown in this study were processed using the first-order planefit command available in the AFM software, which eliminates tilt in the image. Amplitude and phase angle images were processed using the zero-order flatten command, which only modifies the offset of the image.

The schematic diagram of a tip–cantilever assembly in an AFM is shown in Fig. 2.3. The cantilever can scan a sample with its tip in constant contact, intermittent contact, or without contact with the sample surface (Bhushan, 1999a,

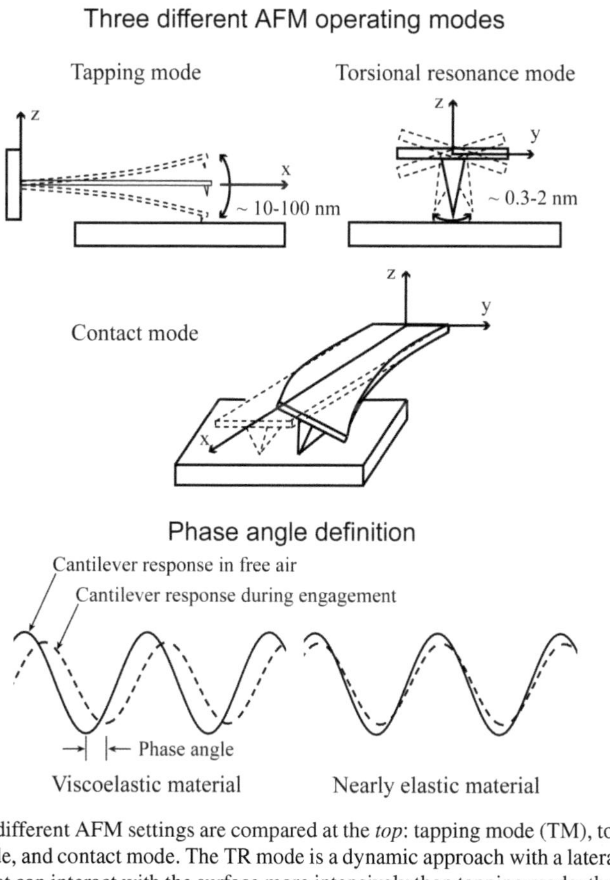

Fig. 2.3 Three different AFM settings are compared at the *top*: tapping mode (TM), torsional resonance (TR) mode, and contact mode. The TR mode is a dynamic approach with a laterally vibrating cantilever tip that can interact with the surface more intensively than tapping mode; therefore, more detailed near-surface information is available. Phase angle is defined and two examples of the phase angle response are shown at the *bottom*: one is for materials exhibiting viscoelastic properties (*left*) and the other nearly elastic (*right*)

b, 2002, 2008b). The scanning is implemented by the motion of a cylindrical piezoelectric tube, which can act as the holder of either the cantilever or the sample. The deflection of the cantilever is generally measured using an optical lever method. A laser beam is projected on and reflected from a location on the upper surface of the cantilever close to the tip and led by a mirror into a four-segment photodiode. The normal and lateral deflections of the cantilever at that location can then be obtained by calibrating the voltage output of the photodiode. AFM measurements can be performed with one of the several modes: tapping mode, torsional resonance mode (TR mode) (Kasai et al., 2004), and contact mode, as shown in Fig. 2.3. The phase angle is also defined in Fig. 2.3 as a phase delay between input/output strain and/or stress profiles. Table 2.2 summarizes the characteristics of tapping mode, torsional resonance modes, and contact mode (Chen and Bhushan, 2005). Contact mode is a static mode and uses a non-vibrating tip; therefore, a phase analysis is not available.

The TR mode measures surface height and phase angle (and amplitude) images as follows. The tip is vibrated in the torsional mode at the resonance frequency of the cantilever beam in air driven by a specially designed cantilever holder. The torsional vibration amplitude of the tip (TR amplitude) is detected by the lateral segments of the photodiode. A feedback loop system coupled to a piezoelectric scanning stage is used to control the vertical z position of the sample, which changes the degree of in-plane (lateral) tip–sample interaction of interest. The z displacement of the sample gives a surface height image of the sample. There are two possible operating modes depending on which parameter is controlled (see Table 2.2):

(1) TR mode I: constant TR amplitude
(2) TR mode II: constant normal cantilever deflection (constant load)

Both modes are operated at the resonance frequency of cantilever in air, which is different from the TR friction mode used in a previous study (Bhushan and Kasai, 2004) in which the tip is vibrated at the resonance frequency of the cantilever after engagement. During the measurement, the cantilever–tip assembly is

Table 2.2 Summary of various operating modes of AFM for surface imaging

Operating modes	Direction of cantilever vibration	Parameter controlled	Data obtained
Tapping	Vertical	Setpoint (Constant amplitude)	Surface height, phase angle (normal viscoelasticity)
TR mode I	Torsional (lateral)	Setpoint (Constant amplitude)	Surface height, phase angle (lateral viscoelasticity)
TR mode II	Torsional (lateral)	Constant load	Surface height, amplitude, and phase angle (lateral stiffness and viscoelasticity)
Contact	n/a	Constant load	Surface height, friction

first vibrated at its resonance at some amplitude before the tip engages the sample. Then, for TR mode I, the tip engages the sample at some setpoint which is reported as a ratio of the vibration amplitude after engagement to the vibration amplitude in free air before the engagement (Bhushan and Li, 2003; Kasai et al., 2004; Scott and Bhushan, 2003).

For TR mode II, instead of keeping a constant setpoint, a constant normal load measured using vertical segments of the photodiode is applied. Under in-plane tip–sample interaction, torsional resonance frequency, amplitude, and phase of the cantilever all change from those when it is far away from the sample surface and could be used for contrasting and imaging of in-plane lateral surface properties.

Compared to TM and TR mode I, the AFM tip interacts with the surface more intensively in TR mode II; therefore, more detailed in-plane surface information can be obtained (Chen and Bhushan, 2005). By using TR mode II, TR amplitude and TR phase angle images show even larger contrast. Previous studies (Bhushan and Li, 2003; Kasai et al., 2004; Scott and Bhushan, 2003) indicated that the phase shift can be related to the energy dissipation through the viscoelastic deformation process between the tip and the sample. Recent theoretical analysis (Song and Bhushan, 2005) has established a quantitative correlation between the lateral surface properties (lateral stiffness and viscoelasticity) of materials and amplitude/phase angle shift in TR measurements. The contrast in the TR amplitude and phase angle images is due to the in-plane (lateral) heterogeneity of the surface. Based on the TR amplitude and phase angle images, the lateral surface properties (lateral stiffness and viscoelasticity) mapping of materials can be obtained. In the work presented in the book, TR mode II amplitude and phase images were obtained to characterize the cellular structure of human hair. For convenience, TR amplitude is recorded in volt, and 1 V corresponds to about 0.5 nm TR amplitude.

2.2.2 Surface Potential Studies Using AFM-Based Kelvin Probe Microscopy

All measurements were taken with a MultiMode atomic force microscope equipped with Extender Electronics modules. The extender allows for surface potential measurements to be taken. Surface potential measurement is conducted using a two-pass method (Bhushan and Goldade, 2000a, b; DeVecchio and Bhushan, 1998; Lodge and Bhushan, 2007a, b; Seshadri and Bhushan, 2008c). In the first pass, surface topography is measured using the standard AFM tapping mode; see Fig. 2.4a. In the second pass, the tip is scanned over the previously measured topography at a specified distance above the surface (for example, 30 nm) – Fig. 2.4b. The piezo normally oscillating the tip in tapping mode is turned off. Instead an oscillating voltage is applied directly to the conducting tip which generates an oscillating electrostatic force. To measure the surface potential, a dc voltage is applied to the tip until the voltage difference between the tip and sample is equal to zero, giving zero oscillating force amplitude. The conducting tip, a silicon tip coated with Pt/Ir or Co/Cr, is used.

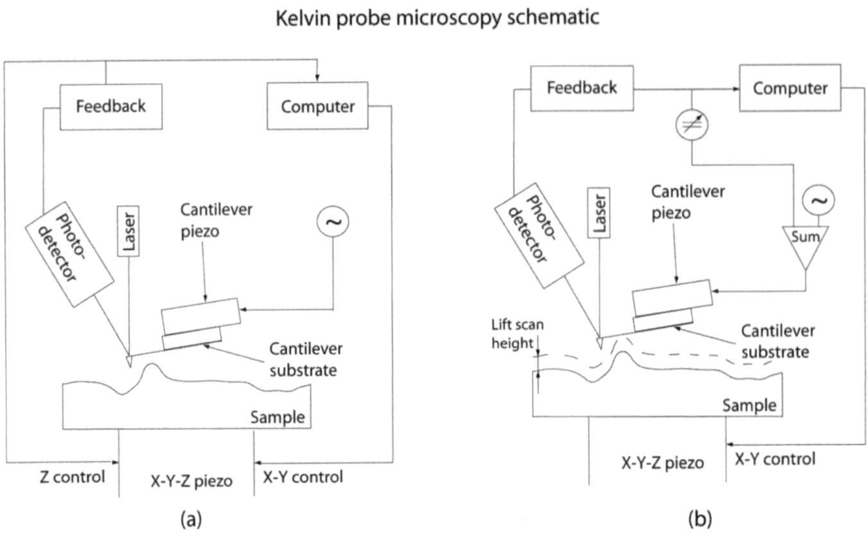

Fig. 2.4 (**a**) Schematic of first pass of Kelvin probe technique measuring surface height; (**b**) schematic of second pass of Kelvin probe technique measuring surface potential (Bhushan and Goldade, 2000b)

Hair samples were mounted in two different ways. When mounted in silver paint, there exists a direct path from the sample to ground. When mounted in liquid paper, the sample is electrically isolated from ground.

2.2.3 Nanomechanical Characterization Using Nanoindentation

2.2.3.1 Nanoindentation

Figure 2.5 shows the schematic of performing nanoindentation and nanoscratch tests on human hair surface with a Nano Indenter II® (MTS Systems Corp.). This instrument monitors and records the dynamic load and displacement of the indenter during indentation with a force resolution of about 75 nN and displacement resolution of about 0.1 nm (Bhushan, 1999b, 2008b; Bhushan and Li, 2003).

For hardness, Young's modulus, and creep measurement, a three-sided (triangular-based) pyramidal diamond Berkovich indenter tip (radius 100–200 nm) was used (Wei and Bhushan, 2006; Wei et al., 2005). For nanoindentation on all the virgin hair surface except African hair, a wide load range (0.1–300 mN) was used, in order to study the mechanical property variation of hair surface depending on the indentation depth. In the case of virgin African hair, the load of 300 mN was not used because the hair was too soft to get reasonable data at 300 mN. For the damaged hair, treated hair, and the hair near scalp, in the middle, and near tip, 0.1, 1.0, 10, and 100 mN normal loads were used. At each load, five indents were made, and the hardness and elastic modulus values were averaged from them, and the standard deviations

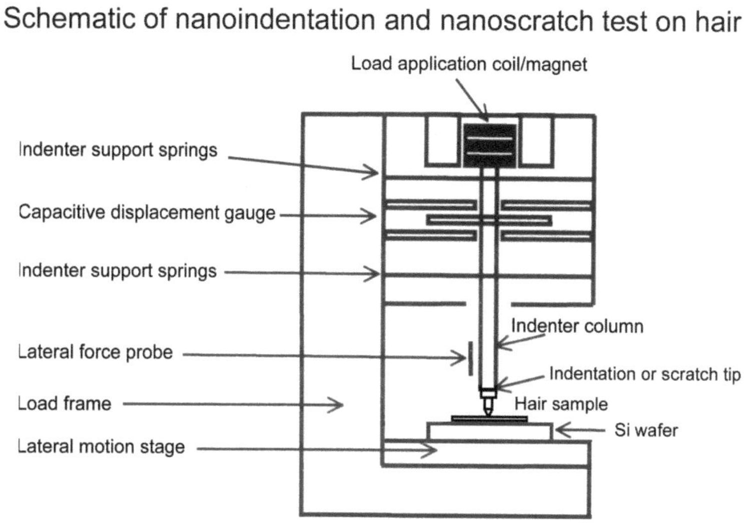

Fig. 2.5 Schematic of nanoindentation and nanoscratch test on hair using nanoindenter

were calculated. For nanoindentation on virgin hair cross-section samples, only one normal load, 1.0 mN, was used to make indents at the cuticle, cortex, and medulla. To do the creep test, a normal load was applied, and then the tip was held for 600 s. The displacement changes during the holding time were recorded. The loads used were 0.1, 1.0, and 10 mN.

2.2.3.2 Nanoscratch

For coefficient of friction and scratch-resistance measurement, a conical diamond tip, having a tip radius of about 1 μm and an included angle of 60°, was used. Before scratching, the hair sample holder was manually rotated so that the hair axis is parallel to the scratch direction.

The scratch tests were performed on both the single cuticle cell and multiple cuticle cells of each hair sample, by controlling the scratch lengths to be 5 μm at a maximum load of 1 mN and 50 μm at a maximum load of 10 mN. Each scratch test was repeated at least five times on the same hair to verify the data reproducibility. For the 50 μm long scratch test, two scratch directions were used to study the directionality effect of the cuticle. One scratch direction was along the cuticle, and the other was against the cuticle. The coefficient of friction was monitored during scratching. In order to obtain scratch depths during scratching, the surface profile of the human hair was first obtained by translating the sample at a low load of about 0.01 mN, which is insufficient to damage the hair surface. The 5 and 50 μm long scratches were then made by translating the hair sample at a constant tip velocity of 0.5 μm/s while ramping the loads from 0.01 to 1 mN and from 0.01 to 10 mN, respectively. The actual depth during scratching was obtained by subtracting the initial profile from the scratch depth measured during scratching. In order to measure the residual depth after scratch, the scratched surface was profiled at a low load of 0.01 mN and was subtracted from the surface profile before scratching. After nanoscratch tests, the scratch wear tracks of the hair samples were measured using a Philips XL-30 ESEM.

In order to study the effect of soaking on hair nanotribological and nanomechanical properties, the hair samples were soaked in de-ionized water for 5 min, which is representative of a typical exposure time when showering/bathing. Then the hair was mounted on a Si wafer immediately, and 5 and 50 μm long scratch tests were performed on them.

2.2.4 In Situ Tensile Deformation Characterization Using AFM

In situ tensile testing of human hair fibers in AFM was conducted using a custom-built stage used in place of the regular sample holder (Bobji and Bhushan, 2001a, b; Seshadri and Bhushan, 2008a, b; Tambe and Bhushan, 2004). It consists of a linear stepper motor which loads a single hair fiber in tension, see Fig. 2.6. The base plate of this stage attaches to the stepper motor of the AFM base, to enable positioning of the stage in X and Y directions with respect to the AFM tip. During scanning, the sample is held stationery while the cantilever tip mounted on an X–Y–Z piezo moves back and forth. The hair sample is firmly clamped in between two sliders to prevent slipping on load application. Stage motion is achieved by a left–right combination lead screw that keeps the sample at approximately the same position with respect to the scanning tip. This helps in locating approximately the same control

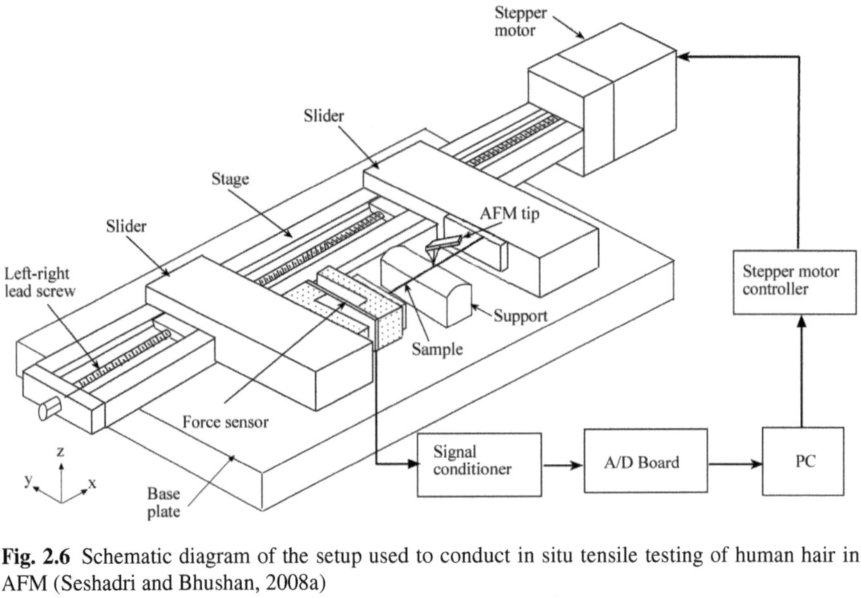

Fig. 2.6 Schematic diagram of the setup used to conduct in situ tensile testing of human hair in AFM (Seshadri and Bhushan, 2008a)

area after each loading increment is applied. With every load increment, there is an increase in the hair length and hence a shift in the location of the control area for the previous position. To locate the same control area in hair, a mark was made on the hair fiber with a nail polish. Because of its reflective nature, it shows up clearly in the AFM optical microscope. A 40 threads per inch (TPI) pitch lead screw in combination with a 400 steps per revolution stepper motor (model PX245–01AA, using the controller NF-90, both from Velmex Inc.) gives a minimum displacement of 1.6 μm. For a sample length of 38 mm (1.5 in.), this corresponds to a minimum strain increment of $8.33 \times 10^{-3}\%$. The strain applied was obtained from the total number of steps through which the stepper motor was rotated. The maximum travel was 10 mm. A beam-type strain gauge force sensor (model LCL-010, Omega Engineering, Stamford, CT), with a resolution of 10 mN, was used for measuring stress in the hair samples. The stiffness of the force sensor (18 kN/m) is very high compared to the sample stiffness. To minimize airborne vibrations during AFM imaging, the hair sample was supported with a smooth aluminum block support having a radius of curvature of 25.4 mm, as shown in Fig. 2.6.

Fatigue experiments were carried out by controlling and cycling strain, through programming the stepper motor of the tensile stage. Cuticle lift-off, the first morphological change during fiber failure, is known to occur at 20% strain (Seshadri and Bhushan, 2008b). Hence the maximum strain during fatiguing was kept below that value at 15% strain. For 15% strain the stress is anywhere between 60 and 80 MPa. The strain rate used during cycling was 38 mm/s. A full cycle took 300 ms. Each fiber was subjected to 5000 such cycles.

2.2.5 Macroscale Tribological Characterization Using a Friction Test Apparatus

The macroscale tribological (friction and wear) characterization of human hair was conducted using a flat-on-flat tribometer under reciprocating motion (Bhushan, 1999a, 2002; Bhushan et al., 2005) (Fig. 2.7). A piece of square polyurethane film (film area 25–400 mm^2) was fixed at the end of a cantilever beam. The hair strands were mounted on a Si wafer in such a way that all strands of hair were separated and parallel to each other. For high-temperature studies, the Si wafer was placed on a heating stage, which can increase the temperature of the hair sample up to 100°C (Bhushan et al., 2005). The heat-generating elements of the heating stage were ohmic resistors encapsulated in a steel holder and kept in good thermal contact by using thermal paste. J-type thermocouples were used to measure the sample temperature. A thermal controller and a solid-state relay were used to control the temperature by adjusting on/off time. A glass plate was attached to the bottom of the heating stage to isolate it thermally from the X–Y axis stage, where the heating stage was mounted. The X–Y axis stage is a lead screw type stage driven by a stepper motor. The load on the polyurethane film was applied by lowering the beam against the hair sample using a microactuator. Normal and frictional forces were measured

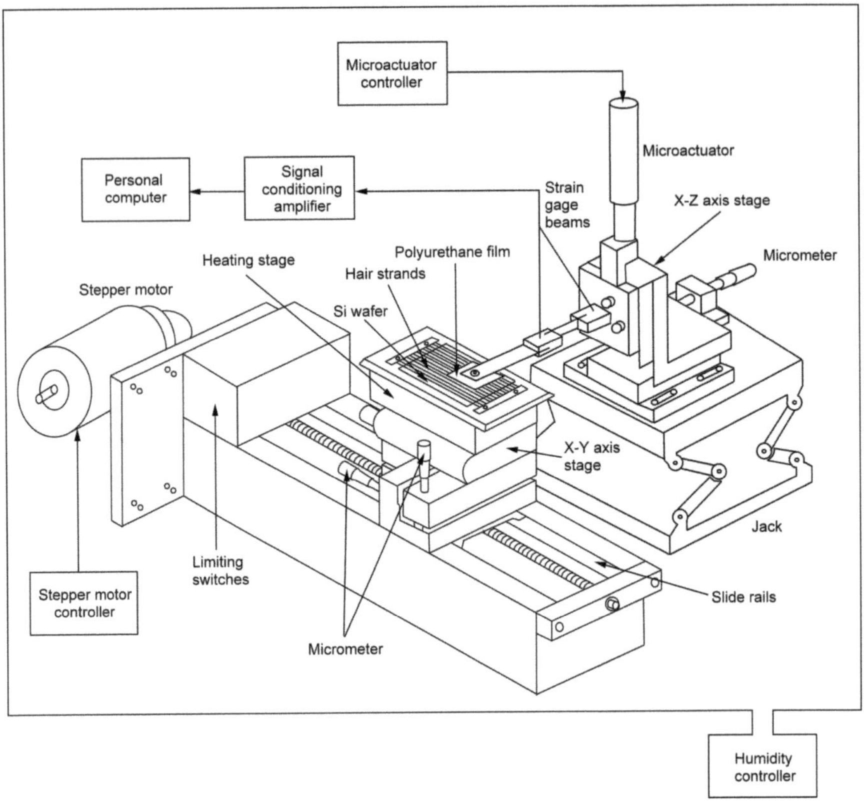

Fig. 2.7 Schematic of the reciprocating tribometer. Normal load is applied by lowering the *X–Z* stage mounted on a laboratory jack. Normal and friction forces are measured by semiconductor strain gages mounted on a crossed-I-beam structure (Bhushan et al., 2005)

with semiconductor strain gage beams mounted on a crossed-I-beam structure as part of the cantilever beam mentioned earlier, and data were digitized and collected on a personal computer. The effect of relative humidity on hair friction was studied in an environmentally controlled chamber, in which humidity was controlled by introducing the combination of dry air and moist air.

In order to select relevant load, velocity, and film area, one needs to be guided by the application. For a feel with a finger with the hair, the normal load applied by the finger was estimated as 50–100 mN, measured by pressing the finger on a microbalance. The estimated apparent contact size and stroke length were 10–100 mm^2 and 5–20 mm, respectively. The sliding velocity was estimated as 5–20 mm/s. To perform a parametric study, tests were performed at a range of operating conditions in the range of interest for the application at a temperature of 22 ± 1°C and 50 ± 5% relative humidity (RH). The following test conditions were used: stroke length 3 mm; sliding velocity 0.4–4.5 mm/s; normal load 50–100 mN; film size 25–400 mm^2. For the friction studies of polyurethane film vs. various hair, hair vs. hair, and polyurethane film vs. virgin and treated hair at dry and wet conditions, the following nominal test conditions were chosen: sliding velocity 1.4 mm/s; normal load 50 mN; and film area 100 mm^2. To simulate the wet conditions, plumber's putty was placed around the stage, and the hair region was filled with water. For the wear measurements, the polyurethane film was rubbed against the virgin and treated Caucasian hair for 24 h at the above nominal conditions, and the coefficient of friction was measured. The hair surface was studied by an optical microscope prior to and after wear tests. To study the effect of temperature and humidity on hair friction, the tests were conducted at the nominal test conditions, and the following parameters were used: temperature 22–80°C and relative humidity 35–85%.

2.2.6 *Micro/nanotribological Characterization Using an AFM*

2.2.6.1 Specimen Mounting

Hair specimens were mounted onto AFM sample pucks using Liquid Paper® correction fluid. A thin layer of the fluid was brushed onto the puck, and when the fluid hardened into a tacky state, the hair sample was carefully placed. The Liquid Paper® dries quickly to keep the hair firmly in place. An optical microscope was used to preliminarily image the specimen to ensure none of the Liquid Paper® was deposited on the hair surface.

Synthetic materials were attached to AFM sample pucks using double-sided adhesive tape.

2.2.6.2 Surface Roughness, Friction Force, and Adhesive Force Measurements

Surface roughness and friction force measurements were performed using a commercial AFM system (MultiMode Nanoscope IIIa, Digital Instruments, Santa Barbara, CA) in ambient conditions (22 ± 1°C, 50 ± 5% relative humidity) (b, 2006; Chen and Bhushan, 2006; LaTorre and Bhushan, 2005a; LaTorre et al., 2006; Lodge and Bhushan, 2006a). For nanoscale measurements, square pyramidal Si_3N_4 tips of nominal 30–50 nm radius attached to the end of a soft Si_3N_4 cantilever beam (spring constant of 0.06 N/m) were used in the contact mode for surface roughness and friction force measurements simultaneously. A softer cantilever was used to minimize damage to the hair. After engagement of the tip with the cuticle surface, the tip was scanned perpendicular to the longitudinal axis of the fiber. The tip was centered over the cross section in order to be at the very top of the fiber, so as to negate the effects caused by the AFM tip hitting the sides of the hair and adding error to the measurements. In order to minimize scanning artifacts, a scan rate of 1 Hz was used for all measurements. Topographical images to characterize the shape and structure of the various hair were taken at 5 × 5, 10 × 10, and 20 × 20 μm^2 scans at a normal load of 5 nN. These scan sizes were suitable for capturing the surface features of multiple scales and scale edges of the cuticle. To characterize roughness, 2 × 2 μm^2 scans of the cuticle surface without edges were used. Friction force

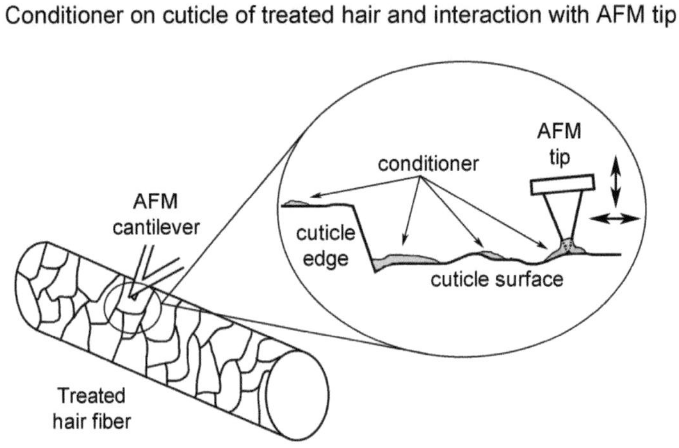

Fig. 2.8 Interactions between the AFM tip and conditioner on the cuticle surface for treated hair (LaTorre and Bhushan, 2005b)

mapping of the scan area was collected simultaneously with roughness mapping (LaTorre and Bhushan, 2005a, b, 2006; LaTorre et al., 2006). Figures 2.2 (bottom cartoon) and 2.8 show the AFM tip scanning over the hair surface for untreated and conditioner treated hair, respectively. The effects of the conditioner can be examined by comparing the friction information. A quantitative measure of friction force was calculated by first calibrating the force based on a method by Bhushan (1999a, 2002, 2008b). The normal load was varied from 5 to 45 nN in roughly 5 nN increments, and a friction force measurement was taken at each increment. By plotting the friction force as a function of normal force, the average coefficient of friction was determined from the slope of the least squares fit line of the data.

Surface roughness images shown in this study were processed using a first-order "planefit" command available in the AFM software, which eliminates tilt in the image. Roughness data as well as friction force data were taken after the "planefit" command was employed. A first-order "flatten" command was also used on friction force mappings to eliminate scanning artifacts and present a cleaner image.

For completeness, it should be noted that Lodge and Bhushan (2006a) have reported some hair damage to occur during imaging in the contact mode. For high-resolution surface height imaging, they used the tapping mode AFM. In this method, a Si tip lightly taps the sample surface at a constant amplitude to gather information on the height of the sample surface. They found that the tapping mode is able to capture more high frequency information than the contact mode due to non-destructive nature.

Adhesive force measurements were made with square pyramidal Si_3N_4 tips attached to the end of a Si_3N_4 cantilever beam (spring constant of 0.58 N/m), using the "force calibration plot technique" (LaTorre 2005a, b, 2006; Chen and Bhushan, 2006; LaTorre et al., 2006; Lodge and Bhushan, 2006a). In this technique, the AFM tip is brought into contact with the sample by extending the piezo vertically, then retracting the piezo, and calculating the force required to separate the tip from the sample. The method is described in detail by Bhushan (1999a, b, 2002, 2008b). The cantilever deflection is plotted on the vertical axis against the Z-position of the piezo scanner in a force calibration plot (Fig. 2.9). As the piezo extends (from right to left in the graph), it approaches the tip and does not show any deflection while in free air. The tip then approaches within a few nanometers (point A) and becomes attached to the sample via attractive forces at very close range (point B). After initial contact, any extension of the piezo results in a further deflection of the tip. Once the piezo reaches the end of its designated ramp size, it retracts to its starting position (from left to right). The tip goes beyond zero deflection and enters the adhesive regime. At point C, the elastic force of the cantilever becomes equivalent to the adhesive force, causing the cantilever to snap back to point D. The horizontal distance between A and D multiplied by the stiffness of the cantilever results in a quantitative measure of adhesive force (Bhushan, 1999a, b, 2002, 2008b).

The force calibration plot allows for the calculation of an adhesive force at a distinct point on the sample surface. Consequently, by taking a force calibration plot at discrete sampling intervals over an entire scan area, a resulting adhesive force mapping (also known as force–volume mapping) can be created to display the variation in adhesive force over the surface (Bhushan and Dandavate, 2000). In this work, plots were taken at 64×64 distinct points over a scan area of $2 \times 2\ \mu m^2$ for all hair types and ethnicities. A custom program coded in Matlab was used to display the force–volume maps. The adhesive force for each force calibration plot was obtained by multiplying the spring constant with the horizontal distance (in the retract mode) traveled by the piezo from the point of zero applied load to the point where the tip snaps off.

For microscale adhesion and friction measurements, a 4 μm radius silica ball was mounted on a Si_3N_4 cantilever (spring constant of 0.58 N/m) (LaTorre and Bhushan, 2006).

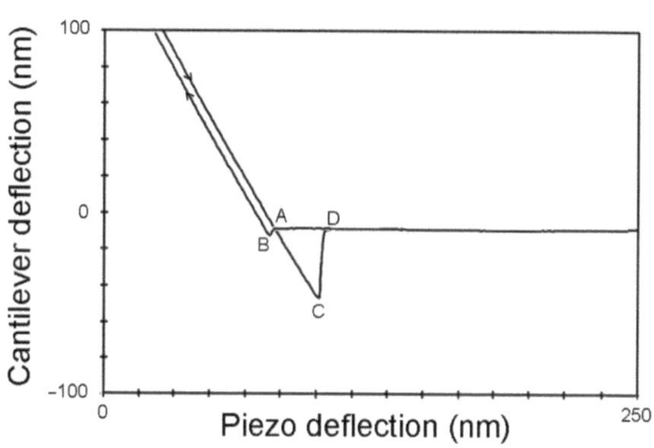

Fig. 2.9 Typical force calibration plot for Caucasian virgin hair. Contact between the tip and hair occurs at point B. At point C, the elastic force of the cantilever becomes equivalent to the adhesive force, causing the cantilever to snap back to point D

2.2.6.3 Relative Humidity, Temperature, and Soaking, and Durability Measurements

A Plexiglas test chamber enclosing the AFM system was used to contain the environment to be humidity controlled. The humidity inside the chamber was controlled by using desiccant to reduce humidity or by adding humid air to increase the humidity. Measurements were taken at nominal relative humidities of 5, 50, and 80%. Hair fibers were placed at each humidity for approximately 2 h prior to measurements.

A homemade thermal stage was used to conduct temperature effect measurements at 20, 37, 50, and 80°C. Hair fibers were exposed to each temperature condition for approximately 30 min prior to testing.

Soak tests were performed as follows: a dry hair fiber was taken from a swatch, and a sample was cut from the fiber (approximately 7 mm long) for coefficient of friction measurements. An adjacent sample was also taken from the fiber and placed in a small beaker filled with de-ionized water. The sample was subjected to the aqueous environment for 5 min, which is representative of a typical exposure time when showering/bathing, then immediately analyzed with an AFM. It should be noted that hair becomes saturated when wet in about 1 min and remains saturated if kept in a humid environment. It was determined from unpublished results that if the wet hair was exposed to ambient environment for more than 20 min while in the AFM, the hair became dry, and the coefficient of friction became similar to that of dry hair. Thus, coefficient of friction measurements were made within a 20 min time frame for each sample.

In order to simulate scratching that can occur on the surface of the hair and its effect on the friction force on the cuticle surface, a durability test was conducted using a stiff silicon tip (spring constant of 40 N/m). A load of 10 μN was used on a 2 μm section of the cuticle. A total of 1000 cycles were performed at 2 Hz. Measurements were conducted using an AFM. Friction force signal was recorded with respect to cycling time.

2.2.6.4 Conditioner Thickness

The measurements were made using AFM with FESP (force modulation etched Si probe) tips (100 nm radius,[1] spring constant of 5 N/m) and using the force calibration plot technique ("force–volume mode" of the nanoscope software). Figure 2.1 shows a schematic diagram of the AFM, and an enlarged cross-sectional view of the AFM tip, conditioner, and hair surface is shown in Fig. 2.10.

The force calibration plot provides the deflection of the cantilever as a function of the piezo position at a distinct point on the sample surface. Figure 2.11 shows a typical force calibration plot curve for commercial conditioner treated hair. From

[1] Typical FESP tip radius is 5–20 nm, but blunt tips were preferred in the study so that when the tip compresses on the surface, the surface tends to deform elastically instead of being indented (plastically deformed). Then the Hertz analysis can be applied to calculate the effective Young's modulus.

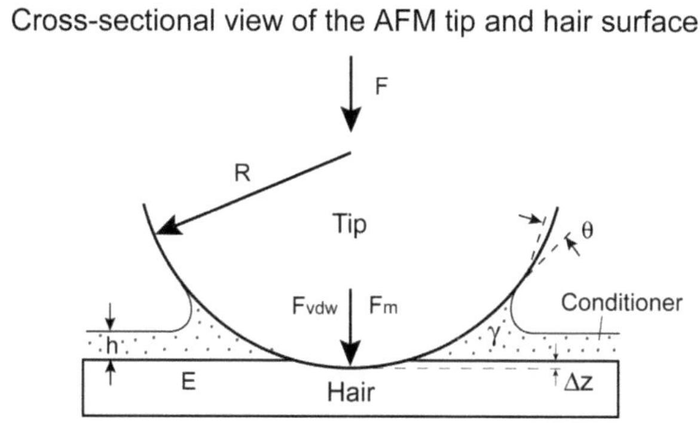

Fig. 2.10 Schematic of an enlarged cross-sectional view of the region around the AFM tip, conditioner, and hair surface. R is the tip radius, h is the conditioner film thickness, γ is the surface tension of the conditioner, θ is the contact angle between the tip and the conditioner, F is the load applied, the adhesive force (F_a) acting on the tip is the sum of van der Waals force F_{vdw} and the meniscus force F_m, Δz is the indention on the hair surface under the applied load F and the adhesive force F_a, and E is the Young's modulus of the sample

this plot, the conditioner thickness and adhesive force can be extracted. The snap-in distance is proportional to the real film thickness (see further details in Sect. 6.1), and the adhesive force F_a (on retract curve), which is the force needed to pull the sample away from the tip and is the sum of van der Waals force F_{vdw} mediated by adsorbed water or conditioner layer and the meniscus force F_m due to Laplace pressure ($F_a = F_{vdw} + F_m$), can be calculated from the force calibration plot by multiplying the spring constant with the horizontal distance between point A and point D in Fig. 2.9. The meniscus force F_m is given by (Fig. 2.10)

$$F_m = 2\pi R\gamma(1 + \cos\theta) \tag{2.1}$$

where R is the tip radius, γ is the surface tension of the conditioner, and θ is the contact angle between the tip and the conditioner.

One limitation with the Young's modulus measurements using a nanoindenter is that it requires loads greater than 1 μN to make accurate measurements. Force calibration plot has been applied to obtain a quantitative measure of the elasticity of compliant samples with an effective Young's modulus as high as a few GPa (Bhushan and Chen, 2006; Chen and Bhushan, 2006). A knowledge of the adhesive forces is necessary for accurate quantitative measurement of the effective Young's modulus. In these measurements, the applied load is much greater than the measured adhesive force so that uncertainties in the adhesive force do not significantly affect the measurements. The Young's modulus of the sample can be determined using Hertz analysis:

$$F + F_a = 4/3\sqrt{R}E\Delta z^{3/2} \tag{2.2}$$

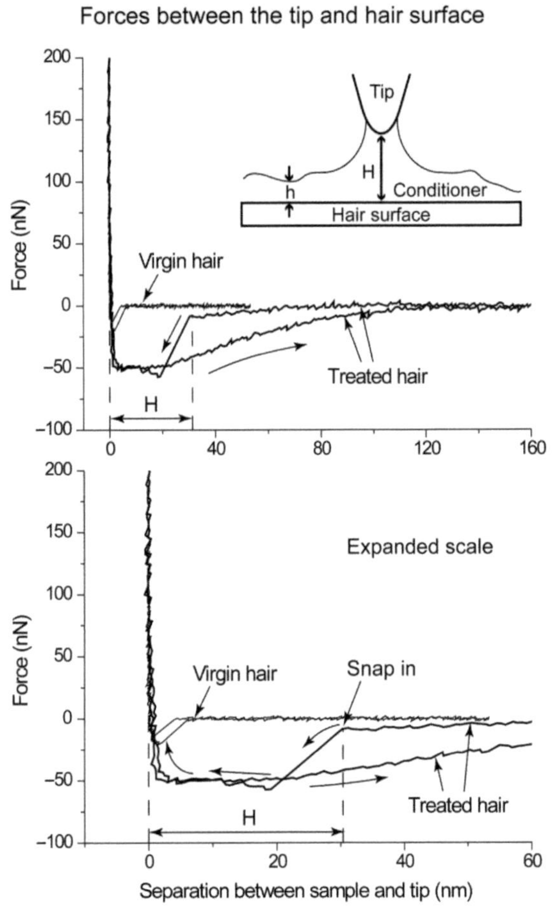

Fig. 2.11 A typical force calibration plot for commercial conditioner-treated hair and schematic diagram of the AFM tip, conditioner, and hair surface

where R is the tip radius, $F + F_a$ is the total force acting on the surface on approach curve which is calculated by multiplying the spring constant with the vertical distance between point C and a point on the line during loading in Fig. 2.9, Δz is the indention on the hair surface, and E is the Young's modulus of the sample. The total force acting on the surface and the resulting deformation (indentation) of the sample Δz can be extracted from the force calibration plot. Additional details are provided in Sect. 6.2.

Consequently, by taking a force calibration plot at discrete sampling intervals over an entire scan area, conditioner thickness, adhesive force, and effective Young's modulus mapping can be created to display the distribution and variation over the surface. In this work, the force curves were collected at the same maximum cantilever deflection (relative trigger mode), at each point in a 64 $\times$ 64 array (total 4096 measurement points) with each force curve sampled at 64 points over a scan area of $2 \times 2\ \mu m^2$ for all hair samples. A custom program coded in Matlab was used to calculate and display conditioner thickness, adhesive force, and Young's modulus maps.

2.3 Hair and Skin Samples

For the research reported in this book, all hair samples were prepared per Appendix A. Samples arrived as hair swatches approximately 0.3 m long. Although the exact location from the root is unknown, it is estimated that hair samples used for testing were between 0.1 and 0.2 m from the scalp. Table 2.3 presents a list of all samples used. The main hair categories of interest are virgin (untreated), virgin (treated with one cycle of commercial conditioner), chemo-mechanically damaged (untreated), chemically damaged (untreated), mechanically damaged (untreated), and chemo-mechanically or chemically damaged (treated with one or three cycles of commercial conditioner or experimental conditioners). Virgin samples are considered to be baseline specimens and are defined as having an intact cuticle and absence of chemical damage. Chemo-mechanically damaged hair fibers were exposed to one or more cycles of coloring and permanent wave treatment, washing, and drying, as well as combing (to contribute mechanical damage), which are representative

Table 2.3 Hair and skin samples

Sample	Type
Caucasian hair	• Virgin • Virgin, treated (1 cycle commercial conditioner) • Chemo-mechanically damaged • Chemically damaged • Chemically damaged, treated (1 cycle commercial conditioner) • Chemically damaged, treated (3 cycles commercial conditioner) • Chemically damaged, treated (PDMS blend silicone or amino silicone) • Mechanically damaged
Asian hair	• Virgin • Virgin, treated (1 cycle commercial conditioner) • Chemo-mechanically damaged
African hair	• Virgin • Virgin, treated (1 cycle commercial conditioner) • Chemo-mechanically damaged
Synthetic materials	• Artificial collagen film (hair) • Polyurethane film (skin) • Human skin (putty replica)

of common hair management and alteration. Chemically damaged hair fibers were not exposed to the combing sequence in their preparation. In the case of African damaged hair samples, chemical damage occurred only by chemical straightening. Mechanically damaged hair fibers were exposed to a combing sequence to cause mechanical damage and were observed under an optical microscope at 100 × to have mechanical damage. All treated hair samples were treated with either one or three rinse/wash cycles of a conditioner similar to a Procter & Gamble commercial product (with polydimethylsiloxane (PDMS) blend silicone), or were treated with two combinations of silicone types (PDMS, blend of low and high molecular weight silicone, or an amino silicone).

Collagen film is typically used as a synthetic hair material for testing purposes. Polyurethane films represent synthetic human skin. They are cast from human skin and have a similar surface energy, which also makes them suitable test specimens

Table 2.4 Contact angle and surface energy of hair and relevant materials associated with nanotribological characterization of hair

	Contact angle (°)		
	Dry	Soaked	Surface energy (N/m)
Virgin Caucasian hair	103[a]	98[a]	0.028[b]
Virgin treated	88[a]	92[a]	
Chemically damaged	70[a]	70[a]	
Chemically damaged, treated	79[a]	84[a]	
Mechanically damaged	95[a]		
Asian	92[a]		
African	80[a]		
PDMS (bulk)	105[c]	0.020[d]	
Human skin – forehead	55[e]		0.043[e]
– forearm	88[e]		0.038[e]
– finger	84[f]		0.029[f]
	74[f]		0.024[f]
	104[g] (after soap washing)		0.027[g]
	58[g] (before shop washing)		
Si_3N_4 tip	48[h]		0.047[i]
Si tip	51[j]		

[a]Lodge and Bhushan (2006b)
[b]LaTorre et al. (2006)
[c]Bhushan and Burton (2005)
[d]Jalbert et al. (1993)
[e]Lerebour et al. (2000)
[f]Schott (1971)
[g]Ginn et al. (1968)
[h]Tao and Bhushan (2006a)
[i]Yanazawa (1984)
[j]Tao and Bhushan (2006b)

when real skin cannot be used. To characterize the surface roughness of human skin, it was also replicated using a two-part silicone elastomer putty (DMR-503 Replication Putty, Dynamold, Inc., Fort Worth, TX). The thickness of the film was approximately 3 mm.

In order to simulate hair conditioner–skin contact in AFM experiments, it is important to have the contact angle and surface energy of an AFM tip close to that of skin. Table 2.4 shows the contact angles and surface energies of materials important to the nanocharacterization of the hair samples: untreated, damaged, and treated human hair; PDMS, which is used in conditioners; skin, which comes into contact with hair; and Si_3N_4 and Si AFM tips used for nanotribological measurements. The dynamic contact angle of various human hair reports in Table 2.4 was measured by Lodge and Bhushan (2006a) using the Wilhelmy balance method. This method uses a microbalance to measure the force exerted on a single fiber when it is immersed into the wetting liquid of interest (de-ionized water). This measured force is related to the wetting force of the liquid on the fiber, and the dynamic contact angle can be calculated. A significant directionality dependence is reported. The values reported here compare well with other values reported in the literature (LaTorre et al., 2006; Molina et al., 2001). The data reported for PDMS (bulk) and live skin in situ were obtained using the standard sessile-drops method using a contact angle goniometer (Bhushan and Burton, 2005; Ginn et al., 1968; Lerebour et al., 2000; Schott, 1971). In the measurement technique developed by Tao and Bhushan (2006a, b) for dynamic contact angle of the AFM, by advancing and receding the AFM tip across the water surface, the meniscus force between the tip and the liquid was measured at the tip–water separation. The water contact angle was determined from the meniscus force.

Chapter 3
Structural Characterization Using an AFM

A schematic diagram in Fig. 1.1a provides an overall view of various cellular structures of human hair. Human hair is a complex tissue consisting of several morphological components, and each component consists of several different chemical species. Table 1.1 presented earlier summarizes the main chemical species present in human hair.

Traditionally, most cellular structure characterizations of human hair or wool fiber were done using SEM and TEM. More recently, the cellular structure of human hair has been characterized using AFM and TR mode II (described earlier) (Bhushan and Chen, 2006; Chen and Bhushan, 2005). In Sect. 3.1, the structure of hair cross section and longitudinal section is presented; and in Sect. 3.2, the structure of various cuticle layers of human hair is presented.

3.1 Structure of Hair Cross Section and Longitudinal Section

3.1.1 Cross Section of Hair

Figure 3.1a shows the AFM images of Caucasian hair cross section. The hair fiber embedding in epoxy resin can be easily seen. From TR amplitude image and TR phase image, the cortex region, the cuticle region, and the epoxy resin region are easily identified. In the cuticle region, five layers of cuticle cells are seen, and the total thickness of the cuticle region is about 2 μm for this sample. In the detailed images of the cuticle region, three layers of cuticle cells are shown, and various sublamellar structures of the cuticle can be seen. The thickness of the cuticle cell varies from 200 to 500 nm. Cortex region shows a very fine circular structure of size about 50 nm, which represents the transverse face of the macrofibril and matrix. At this scale, no intermediate filament structure can be revealed.

3.1.2 Longitudinal Section of Hair

Figure 3.1b shows the AFM images of a longitudinal section of virgin Caucasian hair. Different regions (the cortex region, the cuticle region, and embedding epoxy

B. Bhushan, *Biophysics of Human Hair*, Biological and Medical Physics, Biomedical Engineering, DOI 10.1007/978-3-642-15901-5_3,

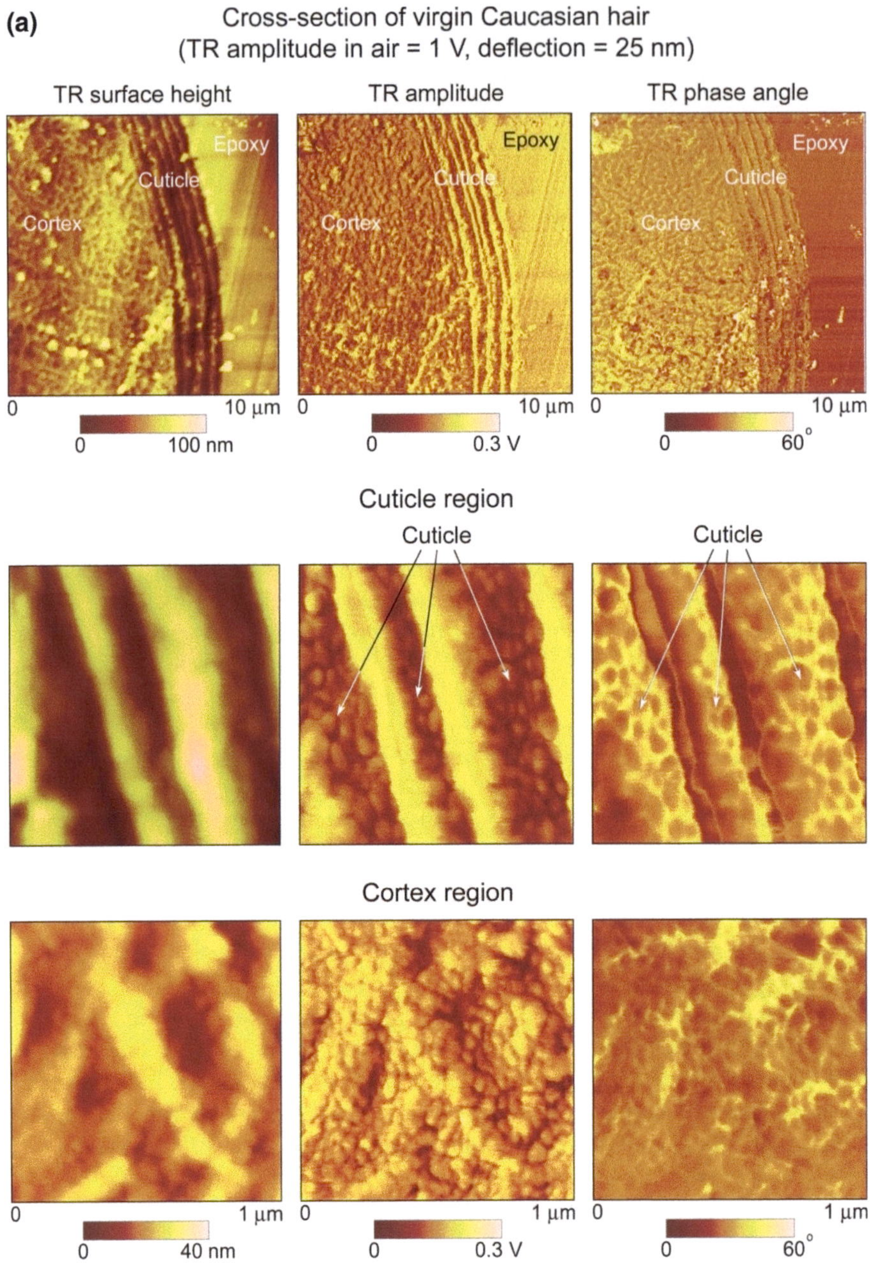

Fig. 3.1 (**a**) Cross-sectional images of virgin Caucasian hair and fine detailed images of the cuticle region and cortex region; (**b**) longitudinal section images of virgin Caucasian hair and fine detailed images of the cuticle region and cortex region. Note that the longitudinal section is not perfectly parallel to the long axis of the hair fiber but with a small angle; therefore, the thickness of sub-lamellar layers of the cuticle is not the real thickness (Chen and Bhushan, 2005)

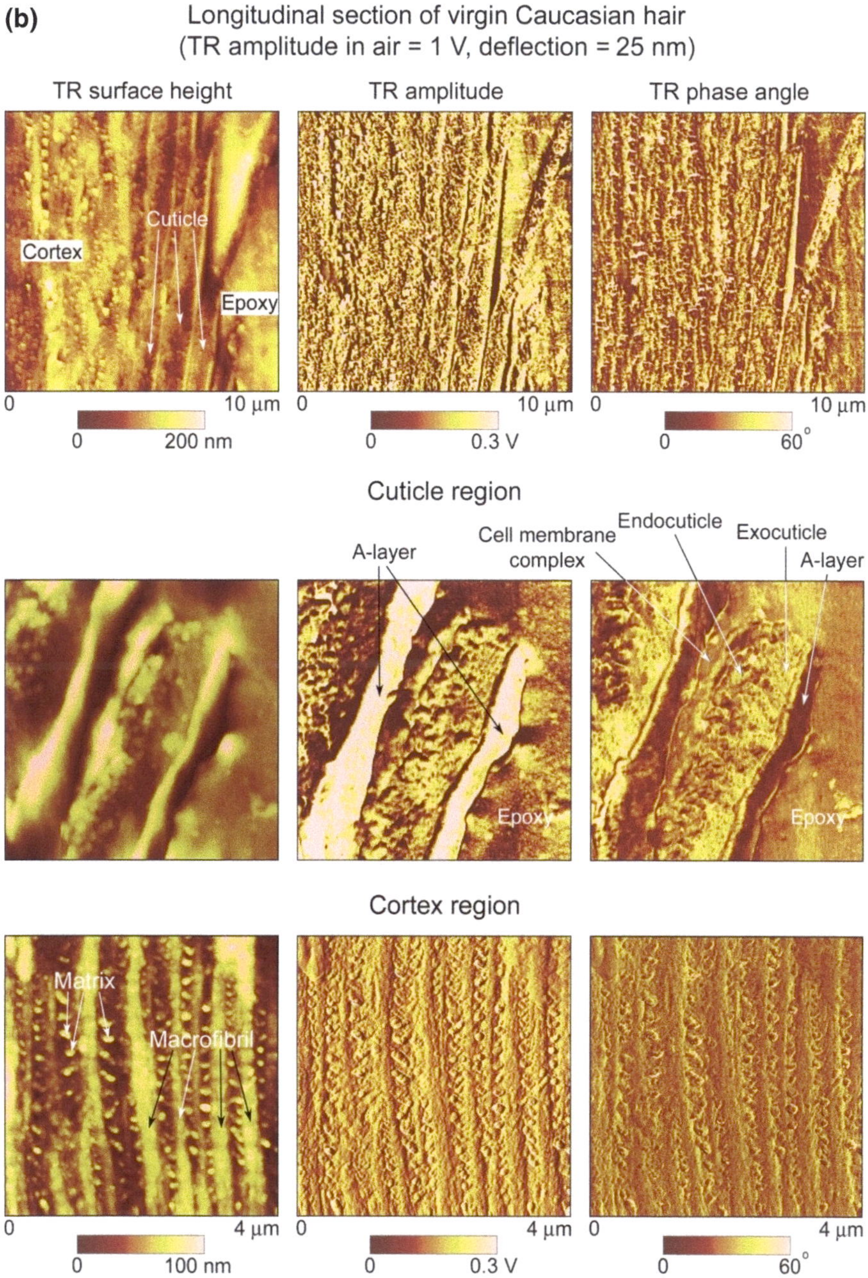

Fig. 3.1 (continued)

resin) are easily seen. As shown in the detailed image of the cuticle region, various sublamellar structures of the cuticle, the A-layer, the exocuticle, the endocuticle, and the cell membrane complex, which cannot be easily revealed in a TR surface height image, are easily resolved in TR amplitude image and TR phase image because of different contrast. Most sublamellar structure features of the cuticle shown in the TEM image of Fig. 1.6b can be identified in the TR amplitude and phase angle images. Previously, these sublamellar structures were only able to be distinguished by TEM (Swift, 1997). Various cellular sublamellar structures in the cuticle have very different chemical content (Robbins, 1994; Swift, 1997): the A-layer is rich in disulfide cross-links due to a very high cystine content of up to 35%; the exocuticle is also rich in disulfide cross-links (15% cystine); in contrast, the endocuticle is relatively lightly cross-linked containing only about 3% cystine. Consequently, these layers exhibit distinct stiffness and viscoelastic properties, and TR mode II imaging technique (TR amplitude and phase angle images) can easily detect these differences. Note that this longitudinal section is not perfectly parallel to the long axis of the hair fiber but with a small angle; therefore, the thickness of various sublamellar layers of the cuticle does not represent the real thickness. In the cortex region, two different morphological regions can be seen: the macrofibril and the matrix. The macrofibril is a bundle of intermediate filaments, aligned parallel to each other and looks like a tree trunk; the matrix surrounds the macrofibril region. The matrix region has a high cystine content compared to the low cystine content of the macrofibril region. This chemical content difference between the macrofibril and the matrix makes it possible to reveal the fine internal cellular structure of hair using AFM TR mode II technique.

3.2 Structure of Various Cuticle Layers

3.2.1 Virgin Hair

Figure 3.2 shows the AFM images of the surface of virgin Caucasian hair. Two typical sample positions are shown: position 1 is near the root end of the hair, and position 2 is near the tip end of the hair. In position 1, one cuticle edge is shown, which is also seen in the TR amplitude and phase images as black strips because of the topographic effect near the cuticle edge. The topographic effect tends to be significant only when there is a large local geometry change. The cuticle edge shows little natural weathering damage and is still intact with a step height of about

Fig. 3.2 Images of surface of virgin Caucasian hair. Two typical samples are shown: near root end in which intact cuticle edges are seen and near tip end in which damage occurs, and part of the cuticle top layers were removed and underneath sublamellar layers are exposed. Detailed images of the outer β-layer, the A-layer, and the endocuticle exposed near tip end are shown at the *bottom* (Chen and Bhushan, 2005)

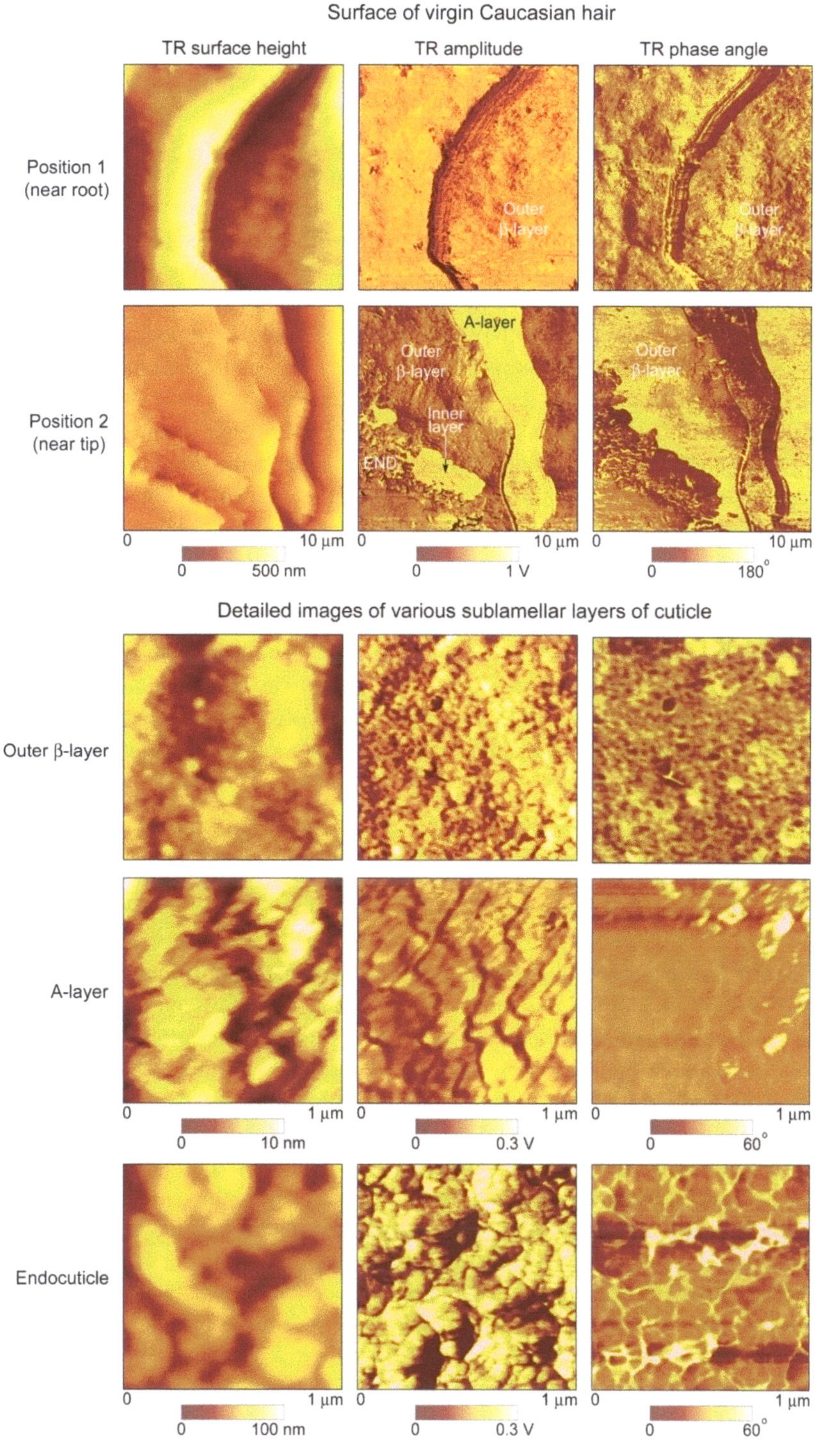

Surface of virgin Caucasian hair
TR surface height
TR amplitude
TR phase angle
Position 1
(near root)
Outer
β-layer
Outer
β-layer
Position 2
(near tip)
A-layer
Outer
β-layer
Inner
layer
END
Outer
β-layer
0
10 μm
0
500 nm
0
10 μm
0
1 V
0
10 μm
0
180°
Detailed images of various sublamellar layers of cuticle
Outer β-layer
A-layer
0
1 μm
0
10 nm
0
1 μm
0
0.3 V
0
1 μm
0
60°
Endocuticle
0
1 μm
0
100 nm
0
1 μm
0
0.3 V
0
1 μm
0
60°

500 nm, and the general cuticle surface, which is covered with a lipid layer (the outer β-layer), is relatively uniform at large scale. In contrast, the surface near the tip of hair (position 2) shows lots of damage, which may be simply because of natural weathering and mechanical damage from the effects of normal grooming actions, such as combing, brushing, and shampooing. Parts of the cuticle outer sublamellar layers were removed, and underneath layers (the A-layer, the endocuticle, and the inner layer) are exposed. Because the different chemical content of various sub-lamellar layers of the cuticle results in different stiffness and viscoelastic properties, a large contrast can be seen in TR amplitude and TR phase angle images. Note that the surface height within each individual sublamellar layer (the A-layer, the outer β-layer, the inner layer) is relativity uniform; therefore, the topographic effect on the TR amplitude and phase is minimal. Detailed images of the outer β-layer, the A-layer, and the endocuticle are shown at the bottom of Fig. 3.2. All these layers show a distinct morphology which can be readily revealed in the TR amplitude and phase angle images: the outer β-layer shows a very fine granular structure; the A-layer shows little discriminable features; and the endocuticle has much rougher granular structure (Swift, 1991, 1997) than that of the outer β-layer. Previous friction force microscopy (FFM) studies (Smith and Swift, 2002) on keratin fibers indicated that for untreated (virgin) fibers, no image contrast was observed on the outer-facing surfaces of the scales. These results indicate that the outer lipid layer may form fine domains, which result in the fine granular structure shown in the TR amplitude and phase images (Fig. 3.2). TR mode II technique has a higher sensitivity compared to the FFM technique; therefore, the fine chemical distribution, which normally cannot be detected by FFM and other techniques, can readily be revealed.

3.2.2 Chemically Damaged Hair

Figure 3.3 shows the AFM images of the surfaces of chemically damaged Caucasian hair. Two typical samples are shown. More damage can be seen compared to the surface of virgin hair. More cuticle edges were removed and often larger areas of the rough granular endocuticle layer were exposed (see sample I). Of the components within each cuticle cell (the A-layer, the exocuticle, the endocuticle, etc.), the endocuticle is the least cross-linked (Robbins, 1994). Under wet conditions it will swell preferentially and is the preferred plane for lamellar fracture under mechanical stress. Indeed, many examples have been observed where the cuticle has come off to leave this granular endocuticle layer of approximately half the thickness of the original scale and located at the scale margins. As shown in sample II, the endocuticle layers were further eliminated, entire pieces of cuticles were removed, and some fine lines on the cuticle surface which delineate the original boundaries of the cuticle edges are clearly seen in TR amplitude images. These lines are referred to as the cuticle edge "ghost" in literature (Smith and Swift, 2002; Swift, 1991). The actual fracture occurs at the interface between the outer β-layer and the δ-layer

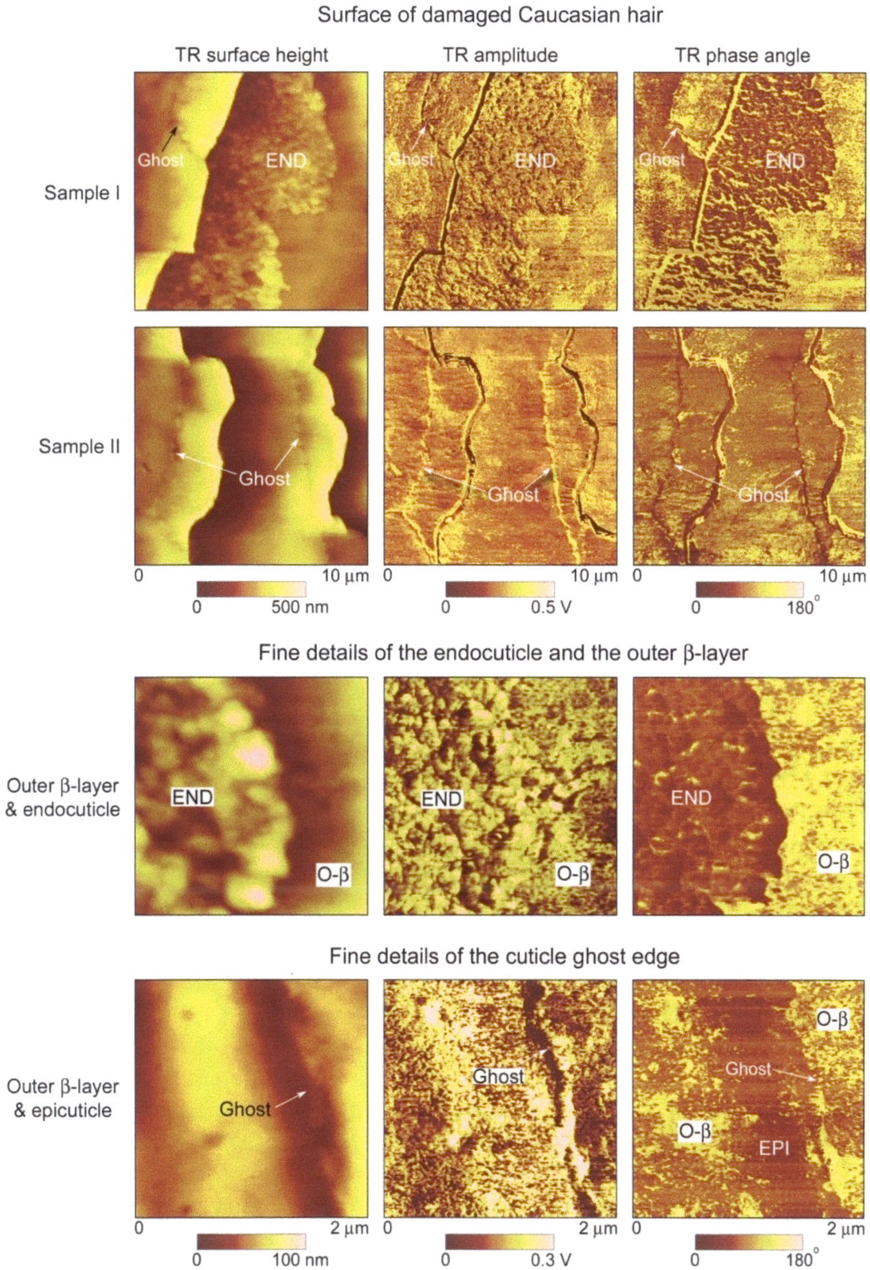

Fig. 3.3 Images of surface of chemically damaged Caucasian hair. Two typical images are shown: sample I in which large areas of the cuticle's top sublamellar layers are removed and much rougher endocuticle layer is exposed; and sample II in which entire pieces of the cuticle are removed and only the cuticle ghost edges are left. Detailed images of the outer β-layer, the endocuticle, and the epicuticle are shown at the *bottom* (Chen and Bhushan, 2005)

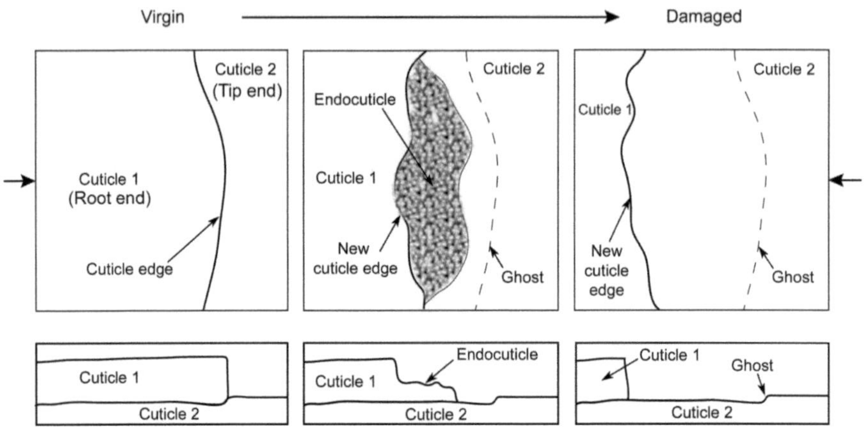

Fig. 3.4 Schematic of the progress of hair damage. The top view images of hair surface at each stage are shown at the *top*. The cross-section views are shown at the *bottom* taken at the corresponding arrows (Chen and Bhushan, 2005)

(see Fig. 1.1). The outer β-layer was originally present, but because of its location, i.e., under the original overlying cuticle but close to the scale edge, it may have undergone oxidative loss through environmental or chemical exposure.

For easy visualization, Fig. 3.4 shows a schematic which illustrates the progress of hair damage. Virgin hair has an intact smooth cuticle edge; as damage occurs (natural weathering or mechanical damage), parts of the cuticle's outer sublamellar layers wear off, and the underneath layers (for example, the endocuticle) are exposed. Further damage will cause an entire piece of cuticle to be broken off, and the ghost which delineates the original boundary of the cuticle edge is seen.

3.2.3 Conditioner-Treated Hair

Various sublamellar layers of the cuticle can be exposed on the surfaces of virgin hair and chemically damaged hair depending on the degree of damage. Because of the distinct chemical nature, these layers may have different interactions with the conditioner (or other hair care products) which will affect the adsorption of the conditioner on the hair surface. Figure 3.5 shows the AFM surface images of two samples of chemically damaged treated hair. In sample I, intact cuticle edges can be seen. From the TR phase angle image, higher contrast can be seen near cuticle edges. It will be presented later that the conditioner is unevenly distributed across the hair surface, and a thicker conditioner layer can be found near cuticle edges (LaTorre and Bhushan, 2005a, b). This buildup of conditioner might be simply caused by the physical entrapment at the steps of the cuticle edges. Uneven thickness of conditioner caused the contrast on TR phase angle images. In sample II, a sharp cuticle edge and a cuticle ghost edge can be readily seen in TR amplitude image. No endocuticle or other sublamellar layers could be found because further treatments

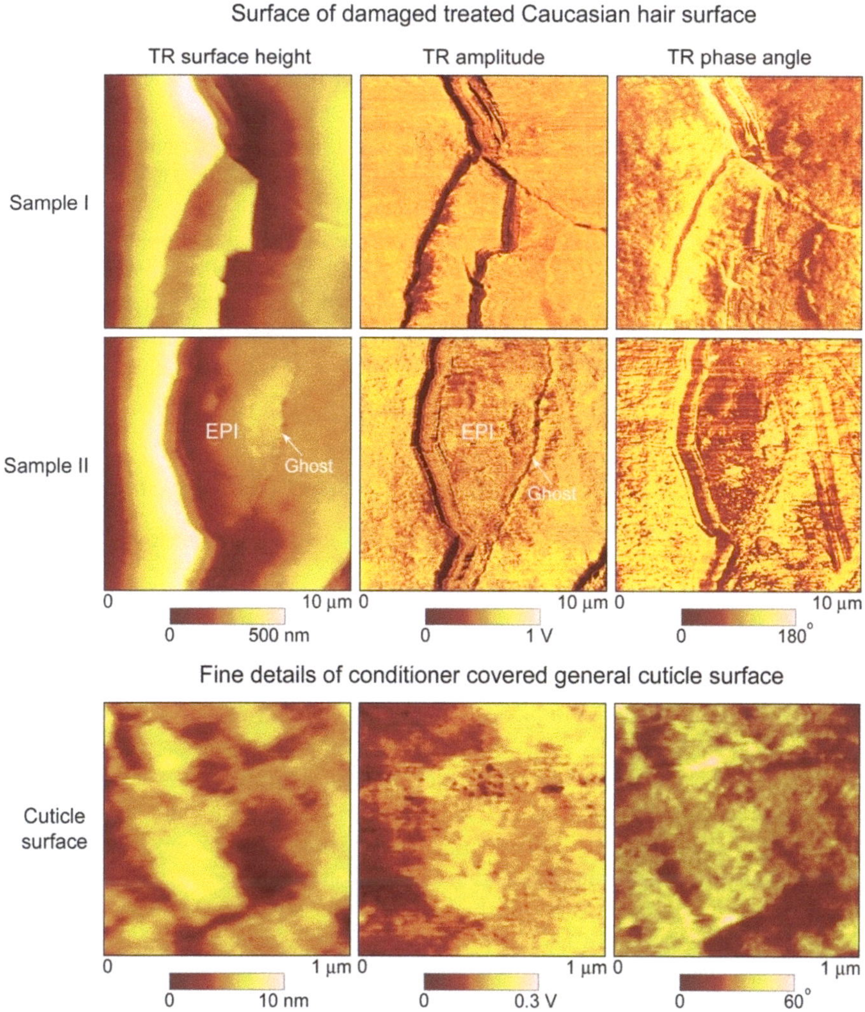

Fig. 3.5 Images of surface of chemically damaged treated Caucasian hair (three cycles of conditioner treatment). The conditioner is unevenly distributed on the hair surface. Thicker conditioner layer can be seen near the cuticle edges. Fine details of conditioner-covered hair surface are shown at the *bottom*. No discernible features can be seen (Chen and Bhushan, 2005)

removed these layers. As shown in the TR phase angle image of sample II, the region between the cuticle edge and the cuticle ghost edge shows different contrast from other parts of hair surface. This newly formed surface may have exposed the epicuticle layer, while the other parts of the hair were still covered with the outer β-layer. The outer β-layer is basically a covalent attached lipid-covered layer which is hydrophobic. The interaction between the conditioner and the outer β-layer or other sublamellar layers of the cuticle (such as the epicuticle) could be very different; therefore, the adsorptions of conditioner on these surfaces are different.

Fine details of chemically damaged treated general cuticle surface (the outer β-layer) are shown at the bottom of Fig. 3.5. Compared to the fine granular structure of the outer β-layer shown in Fig. 3.2, no discriminating features can be seen now since the entire surface is covered with a layer of conditioner.

In summary, although the morphology of the fine cellular structure of human hair has traditionally been investigated using SEM and TEM, these techniques have limited capability to study in situ environmental effects on the physical behavior of hair. AFM TR mode can be used to characterize the fine cellular structure of human hair, and many features previously only seen with SEM and TEM can be identified. The AFM technique provides the possibility for further in situ studies of the effects of environment (temperature, humidity, etc.) and hair care product treatment on the physical behavior of human hair.

3.2.4 *Effect of Humidity on Morphology and Cellular Structure of Hair Surface*

Figure 3.6 shows the TR mode phase contrast images of various hair surfaces at different humidities. The phase contrast images of virgin hair surfaces at different humidities show distinct contrast, indicating a very fine granular structure. Interestingly, this fine granular structure is hardly affected by the humidity changes, as shown in the TR phase contrast images at different humidities. The hydrophobic nature of the intact lipid layer minimizes the effect of water. On the other hand, damaged hair surface shows some contrast at low (5–10%) and medium (40–50%) humidities, but no longer has the fine granular structure as virgin hair surface. Damaged hair surface underwent chemical treatments, which might have partially destroyed the lipid layer on hair surface; therefore, no fine liquid domain can be observed. Some of the inner cellular structures of hair, which are partially hydrophilic, were also exposed. Therefore, a damaged hair surface can adsorb more water at high (80–90%) humidity. In this case, TR mode phase contrast measurement can no longer detect the contrast of different surface domains (content).

Treated hair shows different morphology at different humidities. Treated hair surface is covered with a layer of conditioner, which can adsorb a large amount of water at medium and high humidities. This thick conditioner (and water) layer smears the phase contrast of TR mode images as shown in Fig. 3.6, which is similar to the case of damaged hair at high humidity. However, large phase contrast can be observed at low humidity. At low humidity, the conditioner layer will lose water and is no longer able to retain the water content. The large phase contrast indicates that the conditioner gel network has collapsed after staying at low humidity and formed some kind of isolated conditioner gel network domains. These domains have a distinct chemical content and have different lateral stiffness and viscoelastic properties.

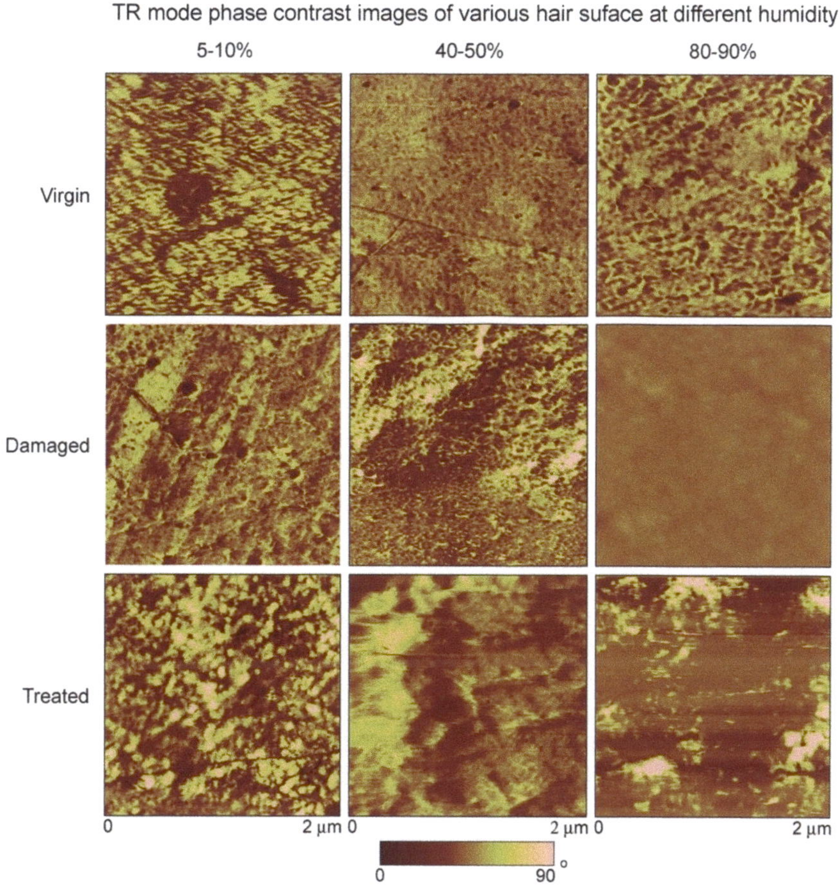

Fig. 3.6 TR mode phase contrast images of Caucasian virgin, chemically damaged, and damaged treated hair surfaces at different humidities (Bhushan and Chen, 2006)

The surface height of a cross section of virgin hair at two humidities is shown in Fig. 3.7. The cortex, layers of the cuticle, and embedding epoxy can be seen. At high humidity, the cortex and cuticle can adsorb a large amount of water, which will consequently weaken the hydrogen and salt bonds between the protein molecules in hair cellular structure.

3.3 Summary

SEM studies of hair cross section and AFM studies of hair surface show that the cuticle is about 5–7 scales thick, and each cuticle cell is about 0.3–0.5 µm thick. The visible length of each cuticle cell is about 5–10 µm long. It appears that the morphology of hair is different from root to tip. That is, the hair near scalp has

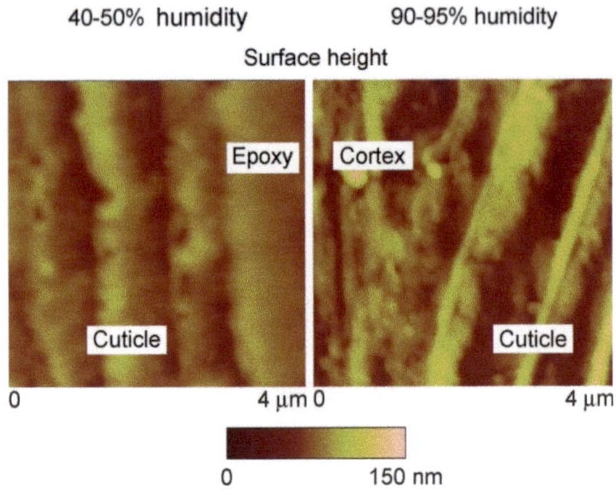

Fig. 3.7 Surface height of virgin Caucasian hair cross section at different humidities (Bhushan and Chen, 2006)

complete cuticles, the hair in the middle has worn cuticles, and the hair near tip seldom has cuticles. The size and shape of Caucasian, Asian, and African hair have been measured from the hair surface and cross section. Asian hair seems to be the thickest (nearly round), followed by African hair (oval-flat), and Caucasian hair (nearly oval).

The cross section and longitudinal section of virgin Caucasian human hair were investigated using TR mode II. The cortex and the cuticle, the macrofibril, and the matrix of human Caucasian hair were readily revealed. Various sublamellar cellular structures of the cuticle, such as the A-layer, the exocuticle, the endocuticle, and the cell membrane complex, are easily observed because of their distinct stiffness and viscoelastic properties. The surface features of various Caucasian human hair (virgin, chemically damaged, and chemically damaged treated) were readily revealed. Sublamellar layers show distinct contrast in TR amplitude and phase angle images. The fine granular structure of the outer β-layer, which has not previously been seen by SEM, TEM, and other AFM studies, is a result of the fine domain formation of lipid layer. Chemically damaged hair surfaces show much more damage; larger areas of the endocuticle were exposed. The endocuticle has a much rougher structure than general cuticle surface (such as the outer β-layer), which could be part of the reason why damaged hair loses shine. Conditioner unevenly distributes on damaged treated hair surface; thicker conditioner films are found near the cuticle edges. At high humidity, the cortex and cuticle can adsorb water, which will consequently weaken the hydrogen bonds and salt bonds between the protein molecules in hair cellular structure. The effect is more significant in chemically damaged and treated hair.

Chapter 4
Nanomechanical Characterization Using Nanoindentation, Nanoscratch, and AFM

Nanomechanical characterization of human hair using nanoindentation and nanoscratch provides valuable information about the hair fiber itself, as well as how damage and treatment affect important mechanical properties of the fiber (Wei and Bhushan, 2006; Wei et al., 2005). In Sect. 4.1, the hardness, Young's modulus, and creep results for both the hair surface and cross section are discussed. In Sect. 4.2, the coefficient of friction and scratch resistance of the hair surface is presented for unsoaked and soaked hair. In Sect. 4.3, stress–strain curves and AFM topographical images of virgin and damaged hair during tensile deformation are presented.

4.1 Hardness, Young's Modulus, and Creep

4.1.1 Hair Surface

Figure 4.1 shows the optical micrograph of three indents on virgin Asian hair made at the normal load of 100 mN. The indentation depths and residual depths were about 5 and 3 μm, respectively. The sizes of these indents are about 15–20 μm in diameter. This image clearly shows that the Nano Indenter II system can successfully make indents on human hair surface.

Figure 4.2a shows the load–displacement curves for virgin, chemo-mechanically damaged, and virgin treated Caucasian hair obtained at two loads: 0.1 and 10 mN. (Loads of 1 and 100 mN were also studied; data not shown.) The hardness and elastic modulus values corresponding to each load–displacement curve are listed in the figure boxes. As mentioned in the experimental section, at each load, five indents were made. Figure 4.2a presents one representative result for each load. At 0.1 mN, the indentation depths of all hair were less than 150 nm, which means that the indents were made within one cuticle scale, assuming that the thickness of one cuticle scale is about 0.3–0.5 μm for all hair samples. At 1.0 mN, the indentation depths were about 0.4–0.6 μm, indicating that indents were probably made through 1–2 cuticle scales. At 10 mN, the indenter tip penetrated about 3–5 cuticle scales. At 100 mN, the indentation depths were about 5 μm, which means that the tip probably reached the cortex of the hair, considering that a hair fiber generally

B. Bhushan, *Biophysics of Human Hair*, Biological and Medical Physics, Biomedical Engineering, DOI 10.1007/978-3-642-15901-5_4,

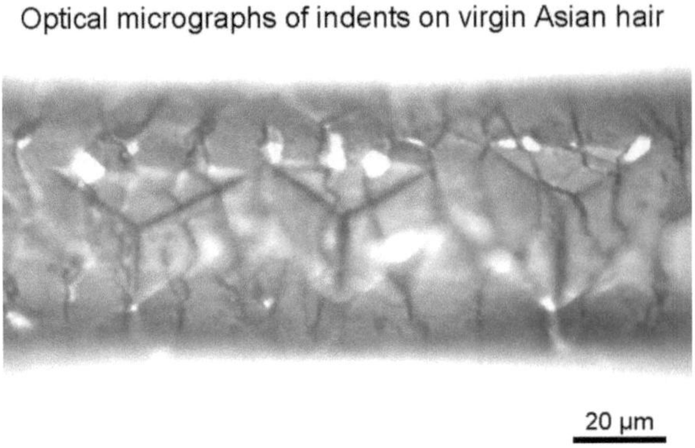

Fig. 4.1 Image of indents on hair surface made using a nanoindenter (Wei et al., 2005)

has about 5–10 cuticle scales. It is interesting to observe that the loading curves of chemo-mechanically damaged hair obtained at 0.1 and 1.0 mN are not as smooth as the virgin and virgin treated hair, especially at the beginning, indicating that the chemo-mechanically damaged hair was very soft at the first 30–50 nm or so, probably because of the chemical damage which caused changes to the exposed surface. At 0.1 and 1.0 mN, the hardness and elastic modulus of virgin treated and chemo-mechanically damaged hair are lower than the virgin hair, indicating that chemical damage and conditioner treatment led to the softness of the hair surface. Considering human hair as a polymeric cylinder, the absorption of chemicals used in hair coloring and conditioner ingredients in the first micron or so might plasticize the polymer and hence reduce its mechanical properties. At 10 and 100 mN, the hardness and elastic modulus of all three hair samples look similar. This result indicates that the effective depth of the chemicals/conditioner influence is probably less than 1.5 μm, i.e., the first 3–4 scales of the cuticles probably interact with the chemicals and conditioner ingredients more effectively than the rest of the scales.

Figure 4.2b shows the hardness and elastic modulus vs. indentation depth for various hair. Every data point (averaged hardness and elastic modulus value), and every error bar in Fig. 4.2b, was calculated from five indentations. According to Fig. 4.2b, the hardness and elastic modulus of hair decreases as the indentation depth increases. In order to explain this, the indentation process is divided into two stages. In the first stage, the indenter tip penetrated the cuticle scales only, in which the indentation depth was probably less than 5 μm. For one cuticle scale, the mechanical properties are expected to decrease from top layer to bottom layer, because the cystine content, and thus the disulfide cross-link density, decreases from the A-layer, to the exocuticle and to the endocuticle (see Fig. 1.1a and Table 1.2). The cuticle scales are bound together by the cell membrane complex, one of the weakest parts of the hair fiber in terms of mechanical properties. The intercellular cement of the cell membrane complex is primarily non-keratinous protein and is low in cystine content (~2%). When the indenter tip penetrated the cuticle scales one by one continuously, the number of the cell membrane complex layers penetrated by

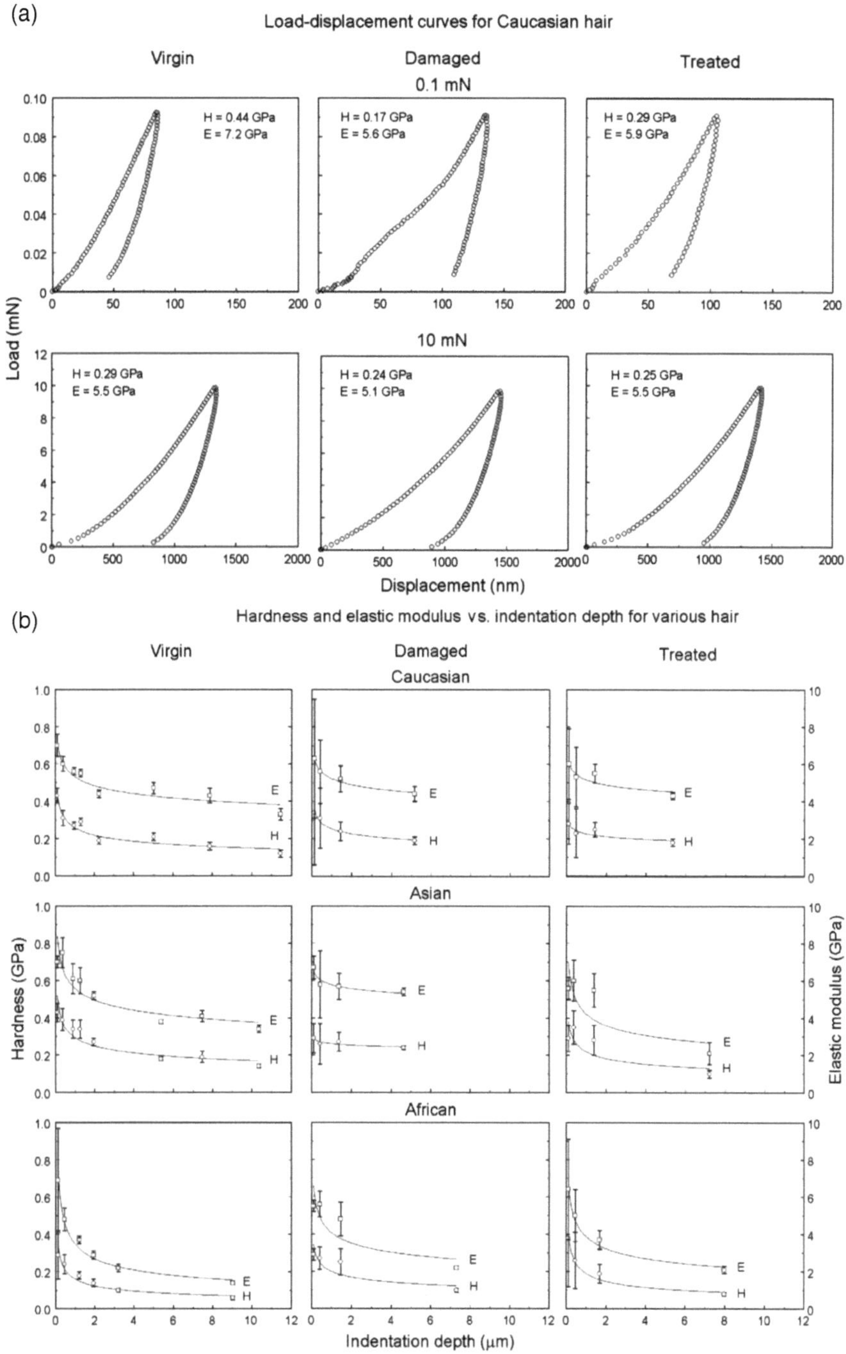

Fig. 4.2 (**a**) Load–displacement curves for Caucasian virgin, chemo-mechanically damaged, and virgin treated hair at two peak loads; (**b**) hardness and elastic modulus vs. indentation depth for various virgin, chemo-mechanically damaged, and treated hair (Wei et al., 2005)

the tip increased. These weak cell membrane complex layers joined together might lead to a deeper displacement upon indentation, contributing to lower mechanical properties. It is also possible that the outer scales of cuticles have higher mechanical properties than the inner scales. In the second stage, the tip began to penetrate the cortex. In general, the cuticles are richer in disulfide cross-links than the cortex (Feughelman, 1997; Robbins, 1994), so the mechanical properties of the cortex are expected to be lower than the cuticle. Putting the two indentation processes together, the hardness and elastic modulus of hair will decrease as a function of the indentation depth.

Figure 4.2b also indicates that at normal loads of 0.1, 1.0, and 10 mN, corresponding to the indentation depth of less than 1.5 μm, the chemo-mechanically damaged and virgin treated hair generally had lower hardness and elastic modulus, but larger error bars, i.e., larger data deviations, than the virgin hair for each ethnicity. This result means that the effective interaction depths were probably less than 1.5 μm for all three ethnicities of hair, and that the effect or distribution of the conditioner on the hair surface was not uniform. It is believed that most of the important interactions between shampoo/conditioner and hair occur at or near the hair surface (the first few micrometers of the fiber periphery). The nanomechanical characterization of hair surface shows that the effective interaction depth ($< 1.5\mu m$) may be shallower than what was thought before. In general, two types of interaction occur between chemical/conditioner ingredients and hair: adsorption and absorption. It has been suggested that for conditioning ingredients in hair conditioners, adsorption is more critical than absorption, because the conditioning ingredients are a relatively large species (Robbins, 1994). If this is the case, then the data variation was probably caused by the non-uniform adsorption of the chemical molecules and the conditioning ingredients to the hair surface. Because the interaction affected the hair up to 1.5 μm deep, absorption should also play an important role here. Transcellular and intercellular diffusion are the two theoretical pathways for absorption to occur. The transcellular route involves diffusion across cuticle cells through both high and low cross-linked proteins. The intercellular diffusion involves penetration between cuticle cells through the intercellular cement and the endocuticle that are low in cystine content (low cross-link density regions). The intercellular diffusion is usually the preferred route for entry of most molecules (especially large ones such as surfactants or even species as small as sulfite near neutral pH). However, for small molecules, transcellular diffusion under certain conditions might be the preferred route, especially if the highly cross-linked A-layer and exocuticle are chemo-mechanically damaged (Robbins, 1994). Depending on the molecular size and the hair condition, the diffusion pathway and diffusion rate might be different from site to site on the hair surface, thus the distribution of conditioner might not be uniform. To sum up, for chemo-mechanically damaged and virgin treated hair, since the adsorption and absorption of chemicals and conditioner ingredients were probably not uniform on the hair surface, the nanomechanical properties of the hair surface (depth $<1.5\mu m$) were not affected (generally decreased) uniformly, leading to the larger data variation compared to the corresponding virgin hair. This implies that the nanoindentation technique can be used to quantitatively evaluate the effective depth of the conditioned hair and distribution of conditioner by measuring the hardness and elastic modulus of the hair surface before and after conditioner treatment as a function of depth and location.

Figure 4.3a summarizes the hardness and elastic modulus of various hair. In general, the chemo-mechanically damaged and virgin treated hair had lower nanomechanical properties and larger error bars than the corresponding virgin hair, as discussed above. The data of African hair was a little strange. For example, the virgin treated African hair seemed to have a higher hardness than virgin African hair. It should be noted that the African hair is naturally curly and highly elliptical, and it was very difficult to mount them and make indentations on their surface. The curly and highly elliptical surface of African hair might cause the indentation results to vary somewhat from the actual values. If the hardness and elastic modulus measured at 1.0 mN is taken as the hair surface hardness and elastic modulus, then by comparing the virgin Caucasian, Asian, and African hair, it is seen that the Asian hair has the highest hardness (0.39 ± 0.06 GPa) and elastic modulus (7.5 ± 0.8 GPa), followed by Caucasian hair with hardness of 0.31 ± 0.04 GPa and elastic modulus of 6.0 ± 0.4 GPa. The African hair seems to have the lowest mechanical properties, whose hardness is 0.24 ± 0.05 GPa and elastic modulus is 4.8 ± 0.6 GPa. Note that all these mechanical properties were measured in the middle part of the hair.

Figure 4.3b summarizes the hardness and elastic modulus for virgin Caucasian hair at three locations: near scalp, middle, and near tip. As expected, the hardness and elastic modulus of the hair surface decreases from root to tip, because of the cuticle damage. Considering that the hair near the scalp has complete cuticles, while the hair near the tip only has exposed cortex, it is probably a good way to compare the nanomechanical properties of the hair cuticle and cortex in the lateral direction by comparing the nanomechanical properties of hair near the scalp and hair near the tip. At 1.0 mN, the cuticle (hair near the scalp) has higher hardness (0.6 ± 0.29 GPa) and elastic modulus (8.4 ± 1.2 GPa) than the cortex (hair near the tip), whose hardness is 0.3 ± 0.06 GPa and elastic modulus is 6.0 ± 0.6 GPa. This result clearly suggests that the cuticles contribute more to the hair lateral mechanical properties than the cortex, which is in good agreement with the theoretical models for wool fibers (Wortmann and Zahn, 1994).

Figure 4.4 shows the creep displacement vs. time curves for various hair. The normal load used for creep tests was 10 mN. In all cases, the displacement increased as time passed. The creep behavior of hair may arise from several sources. Hair is rich in peptide bonds and the abundant CO- and NH-groups present give rise to hydrogen bonds between groups of neighboring chain molecules (Fig. 1.1b). Other linkages, such as side-chain interactions of the disulfide type, and chain folding may also be present in hair. When hair was compressed, the creep behavior was a result of deformation and relaxation of the chemical bonds, the polypeptide chains, and the non-crystalline regions (Barnes and Roberts, 2000). It should be noted that at a normal load lower than 10 mN, the creep behavior was not obvious. Assuming that

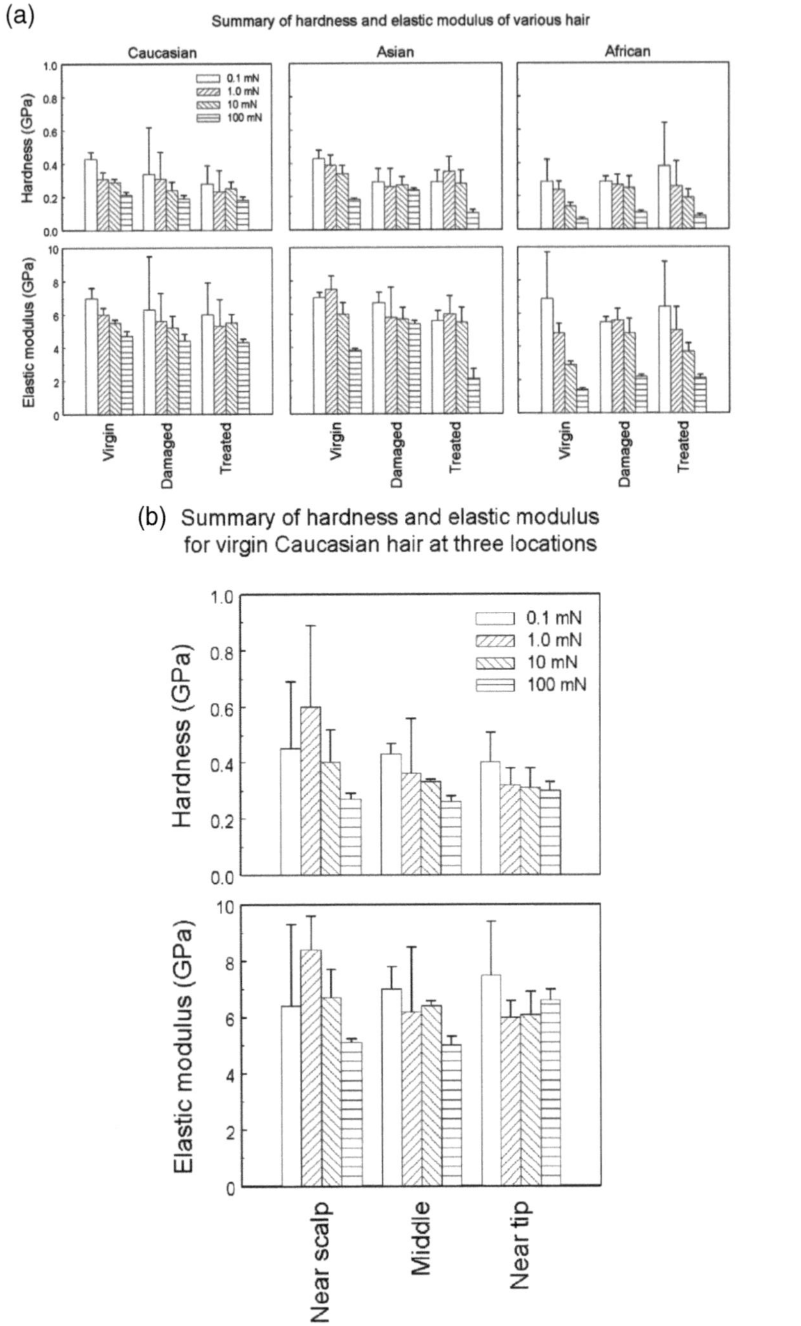

Fig. 4.3 (**a**) Summary of hardness and elastic modulus of virgin, chemo-mechanically damaged, and virgin treated hair at four peak loads; (**b**) summary of hardness and elastic modulus for virgin Caucasian hair at three locations at four peak loads (Wei et al., 2005)

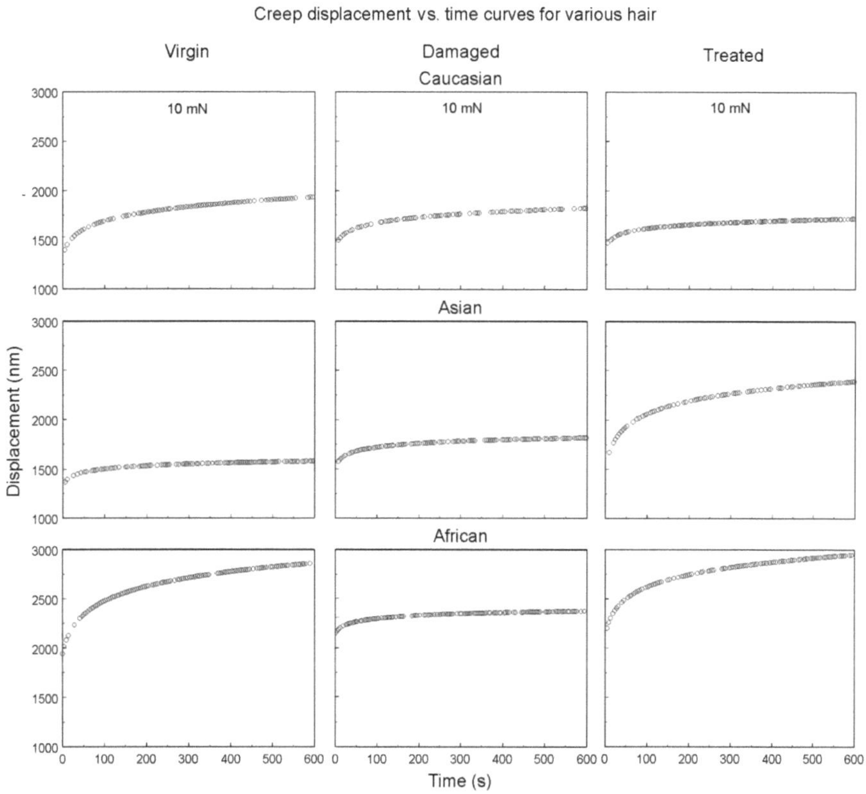

Fig. 4.4 Creep displacement vs. time curves for virgin, chemo-mechanically damaged, and virgin treated hair (Wei et al., 2005)

the diameter of Caucasian, Asian, and African hair was about 50, 100, and 80 μm, respectively, the compression ratio of the indented area of these hair at 10 mN at the beginning of the creep tests were as follows: Caucasian ~2.6%, Asian ~1.3%, and African ~2.5%. This may suggest that if the local compression ratio was less than these values for corresponding hair, the deformation and relaxation of the chemical bonds, the polypeptide chains, and the non-crystalline regions might be too small to cause the creep behavior to occur. According to the creep displacement vs. time curves, it is difficult to correlate the creep behavior of each hair with its ethnicity and condition (virgin, chemo-mechanically damaged, or virgin treated).

4.1.2 Cross Section

Figure 1.6a, presented previously, shows the SEM images of virgin hair cross section. These SEM images may represent the typical shape of Caucasian (nearly oval),

Asian (nearly round), and African (oval-flat) hair. Regarding the diameter, the Asian hair seems to be the thickest, followed by African and Caucasian hair, which is in good agreement with the SEM studies of the hair surface. The center column of Fig. 1.6a shows the cortex and medulla of each hair. The arrows point to the indents made at the cortex. The medulla of African hair is not so obvious, and it is believed that not all the hair has medulla (Robbins, 1994). The right column shows the images of the cuticles. The top-right image clearly shows that the cuticle of Caucasian hair is about 6–7 scales thick, and each cuticle cell is about 0.3–0.5 μm thick. Note that the cuticle scales were separated due to polishing, implying that the binding strength of the cell membrane complex between the cuticle scales might not be very strong.

The hardness and elastic modulus of the hair cuticle, cortex, and medulla were measured from the cross-section samples, and Fig. 4.5 shows the hardness and elastic modulus plots across various virgin hair. As expected, the cuticles have the highest mechanical properties, followed by cortex and medulla. Table 4.1 summarizes the hardness and elastic modulus of various hair (Wei et al., 2005). The hardness of cuticles was taken from the hair surface measurements (see Fig. 4.3a). By comparing the mechanical properties of Caucasian, Asian, and African hair cortex, it can be seen that the Asian cortex appears to have the highest properties, followed by Caucasian and African hair. This trend is in agreement with the trend for the hair surface measurement results (see Fig. 4.3a). Table 4.1 shows that the hardness of cuticles is greater than that of cortex, but the elastic modulus of cortex is comparable to that of cuticle (Wei et al., 2005). Comparing the hardness (0.3 ± 0.06 GPa) and elastic modulus (6.0 ± 0.6 GPa) of Caucasian cortex in the lateral direction (see Fig. 4.3a) with its hardness (0.27 ± 0.02 GPa) and elastic modulus (6.5 ± 0.5 GPa) in the longitudinal direction, it can be seen that hardness and elastic modulus of the hair cortex in the longitudinal direction are lower and higher, respectively, than in the lateral direction.

4.1.3 Effect of Humidity and Temperature on Young's Modulus

Figure 4.6a shows effective Young's modulus mappings of chemically damaged treated hair at different humidities obtained from force calibration plots (Bhushan and Chen, 2006). At different humidities, the maps are distinctly different. As humidity increases, the effective Young's modulus of hair surface decreases significantly. The effective Young's modulus of hair surface at 80–90% humidity is only half the value at 5–10% humidity. At high humidity, the water content in the conditioner gel network will increase significantly; therefore, the conditioner layer on hair surface is thicker and softer, resulting in a lower value of effective Young's modulus. At low humidity, the conditioner gel network may desorb most of the water content, which results in the collapse of the gel network. Consequently, the conditioner layer no longer acts as a soft protection layer, but as a hard, thin shell on the hair surface.

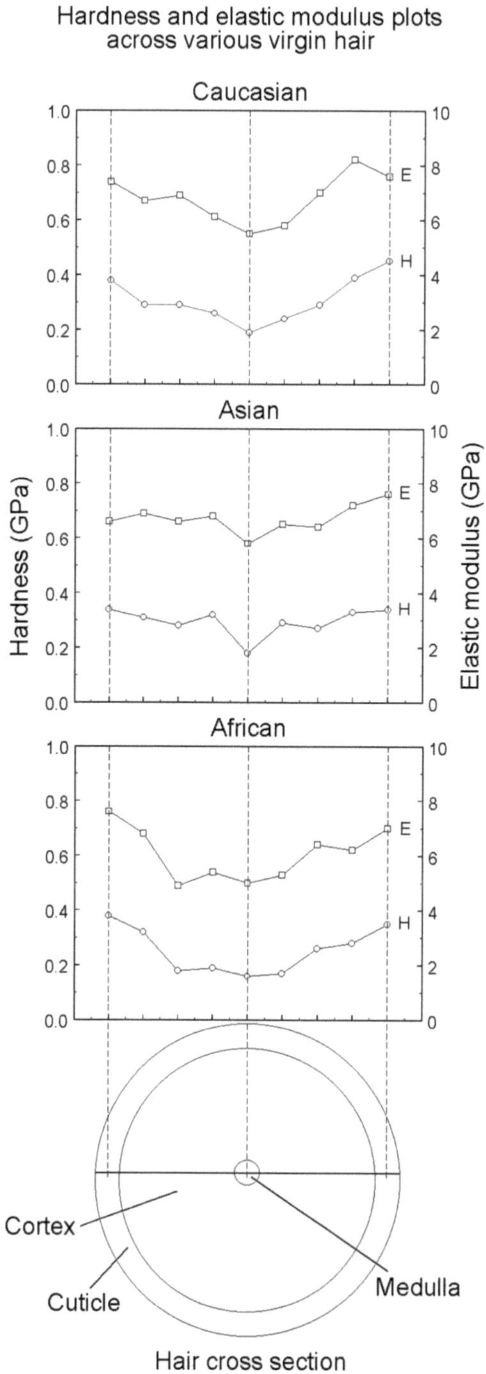

Fig. 4.5 Hardness and elastic modulus plots across various virgin, chemo-mechanically damaged, and virgin treated hair (Wei et al., 2005)

Table 4.1 Summary of hardness and elastic modulus of human hair. Mean and $\pm 1\sigma$ values are presented

	Hardness (GPa)			Elastic modulus (GPa)		
	Cuticle[a]	Cortex[b]	Medulla[b]	Cuticle[a]	Cortex[b]	Medulla[b]
Caucasian	0.32 ± 0.04	0.27 ± 0.02	~ 0.19	6.0 ± 0.4	6.5 ± 0.5	~ 5.5
Asian	0.39 ± 0.06	0.30 ± 0.02	~ 0.18	7.5 ± 0.8	6.7 ± 0.3	~ 5.8
African	0.24 ± 0.05	0.23 ± 0.06	~ 0.16	4.8 ± 0.6	5.8 ± 0.7	~ 5.0

[a]Obtained from the hair surface at normal load of 1.0 mN
[b]Obtained from the hair cross section at normal load of 1.0 mN

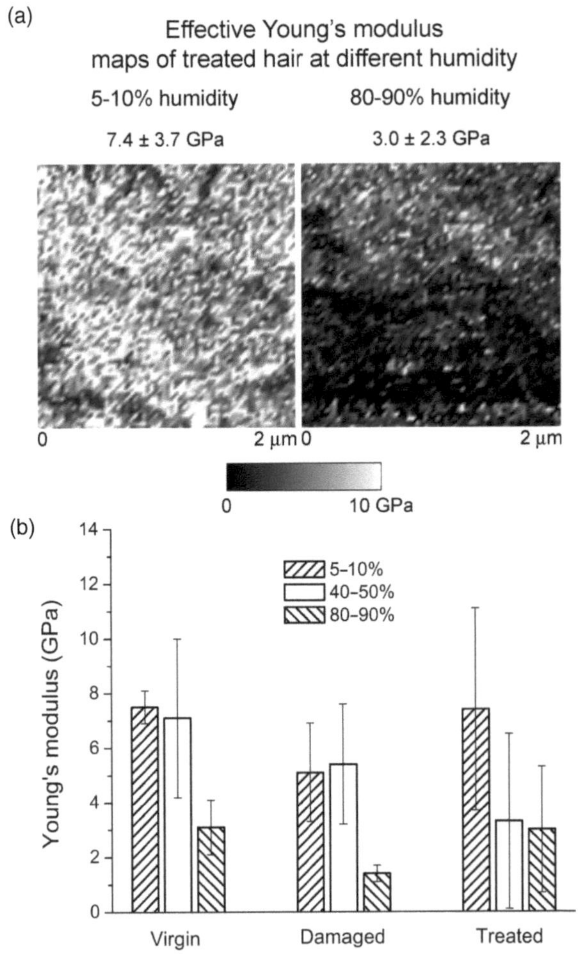

Fig. 4.6 (**a**) Effective Young's modulus maps of Caucasian chemically damaged treated hair at different humidities (Bhushan and Chen, 2006) and (**b**) effective Young's modulus of Caucasian virgin, chemically damaged, and damaged treated hair samples at different humidities (Bhushan and Chen, 2006)

Figure 4.6b summarizes the effective Young's modulus of various hair surfaces at different humidities. Young's modulus of damaged hair is lower than virgin hair, and it decreases dramatically at high humidity. Since the hydrophobic lipid protection layer of damaged hair has been depleted or damaged, the hydrophilic molecules of the inner cellular structure of hair are exposed to water. Water can adsorb and diffuse easily into hair via the defects on the surface, therefore softening the hair. Treated hair has smaller effective Young's modulus than virgin hair at 50% humidity because of the soft physisorbed conditioner gel network layer on the surface. The layer remains intact at high humidity and protects the hair surface from excess water adsorption and diffusion (penetration). However, at low humidity, the gel network is no longer able to retain the water content. It will lose most of the water content and collapse and behave as a hard shell on hair surface. Therefore, treated hair surface at low humidity has the same effective Young's modulus as that of virgin hair.

The effective Young's modulus of the cross section of virgin hair is shown in Fig. 4.7. As humidity increases, the Young's modulus decreases, and the differences between various layers (the cortex, the cuticle, and the epoxy) disappear. Human hair consists of various chemical and physical bonds (salt bond, hydrogen bond and disulfide bond). The strength of these bonds will be strongly affected by the level of water content.

The effective Young's modulus of various hair surfaces was also measured at different temperatures. Three different temperatures are studied: room temperature, human body temperature, and a high temperature which represents the temperature under direct sunshine. Figure 4.8 summarizes the findings for virgin, damaged, and conditioner-treated hair at those temperatures. For virgin or damaged hair surfaces, temperature has little effect on the Young's modulus due to the small range of temperature studied. The effective Young's modulus of conditioner-treated hair surface increases as the temperature increases. At high temperature, the conditioner layer is no longer able to retain the water content and collapses to form a hard shell covering the surface. Therefore, conditioner-treated hair has high effective Young's modulus at high temperature.

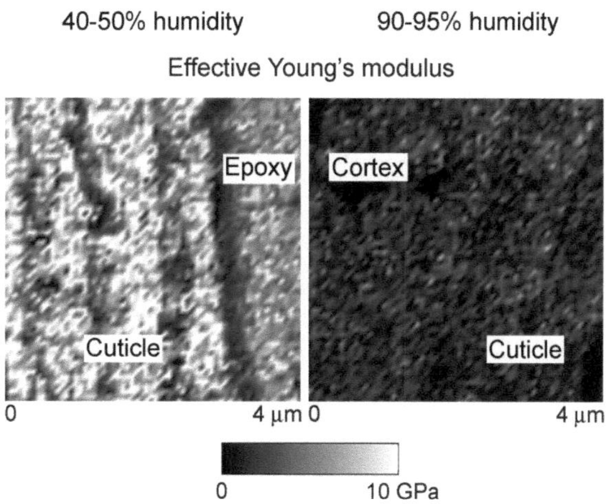

Fig. 4.7 Effective Young's modulus of virgin Caucasian hair cross section at different humidities (Bhushan and Chen, 2006)

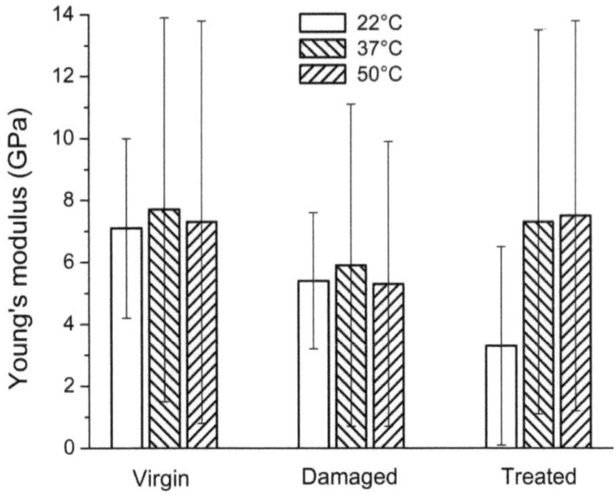

Fig. 4.8 Effective Young's modulus of Caucasian virgin, chemically damaged, and damaged treated hair samples at different temperatures (Bhushan and Chen, 2006)

4.2 Scratch Resistance

Nanoscratch technique is capable of simulating the scratch phenomena on hair surface on the nanoscale by scratching the hair surface using a conical diamond tip (radius about 1 μm) and recording the coefficient of friction, in situ scratch depth, and residual depth.

4.2.1 Nanoscratch on Single Cuticle Cell

Figure 4.9 shows the coefficient of friction and scratch depth profiles as a function of normal load and tip location on a single cuticle cell of Caucasian and Asian hair (virgin, chemo-mechanically damaged, and virgin treated). The scratch direction was from left to right. The scratch length was 5 μm, and the normal load was increased from 0.01 to 1 mN during scratching. The coefficient of friction of all the hair samples ranged from 0.3 to 0.6 (Wei and Bhushan, 2006). The coefficient of friction of virgin treated Caucasian hair (~0.3) is lower than virgin Caucasian hair (~0.4), and the coefficient of friction of virgin treated Asian hair (~0.3) is also lower than virgin Asian hair (~0.5). The conditioner acts as a thin layer of lubricant on hair surface and reduces the coefficient of friction of hair during scratch. The coefficient of friction of chemo-mechanically damaged hair depends on the type and extent of damage. If the chemical damage softens the hair surface, then during scratching, the tip plows into the hair easily, leading to a higher coefficient of friction. If the damage hardens the hair surface or does not change the mechanical properties of the hair surface, then the coefficient of friction probably will decrease or stay the same during scratch.

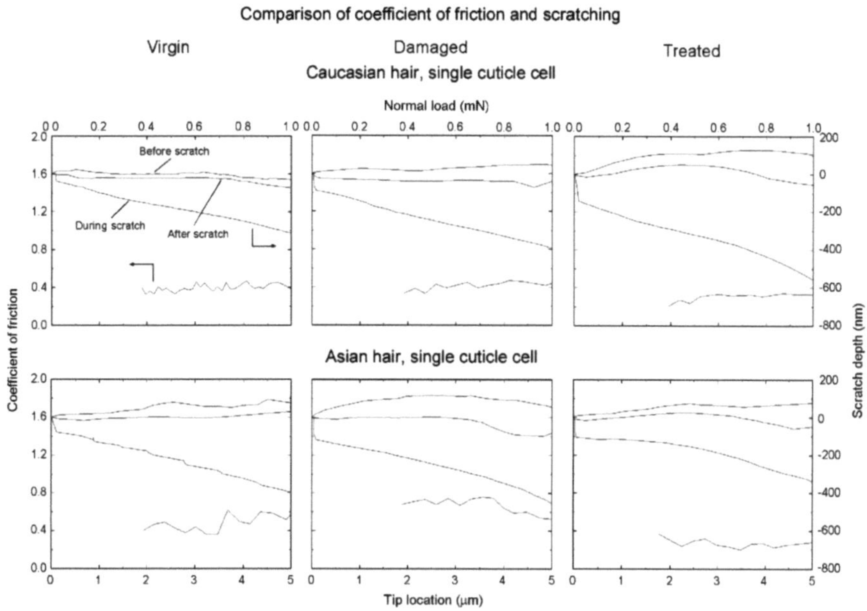

Fig. 4.9 Coefficient of friction and scratch depth profiles as a function of normal load and tip location on single cuticle cell of Caucasian and Asian hair (virgin, chemo-mechanically damaged, and virgin treated) (Wei and Bhushan, 2006)

The scratch depth profiles include the profiles obtained before (pre-scratch), during (in situ scratch), and after (post-scratch) scratching at scratch length of 5 μm and a maximum normal load of 1 mN as indicated in Fig. 4.9. Reduction in scratch depth is observed after scratching as compared to that during scratching. This reduction in scratch depth is attributed to an elastic recovery after removal of the normal load. The post-scratch depth indicates the final depth, which reflects the extent of permanent damage and plowing of the tip into the hair surface. The scratch depth profiles show that at the very beginning of the scratch for all the hair samples, the in situ displacement (30–200 nm) increased rapidly at very low load. After that, it increased gradually. This observation indicates that the top, about 200 nm, of the hair surface may be softer than the underlying layer. The scratch depth profiles also show that the reference surface profile before scratch is not very flat, indicating that human hair has a rough surface. AFM studies have shown that the root mean square (RMS) roughness of Caucasian and Asian hair surface ranges from 7 to 48 nm (LaTorre and Bhushan, 2005a). At 1 mN, the in situ scratch depths of all hair samples range from 300 to 600 nm, and the residual depths range from 50 to 200 nm. Since the thickness of one cuticle cell is about 300–500 nm, during the 1 mN nanoscratch test, the scratch tip might only penetrate one cuticle cell layer.

4.2.2 Nanoscratch on Multiple Cuticle Cells

Most of the time when we comb our hair, the comb is scratching multiple cuticle cells. Figure 4.10a shows the coefficient of friction and scratch depth profiles as a function of normal load and tip location on multiple cuticle cells of chemo-mechanically damaged Caucasian hair obtained in two scratch tests: scratch along the cuticle and scratch against the cuticle; and Fig. 4.11b shows the SEM images of the hair surface after scratch. The coefficient of friction obtained when the tip scratched the hair against the cuticle is significantly higher than the coefficient of friction obtained when the tip scratched the hair along the cuticle, which is known as the "directionality effect." This is understandable because when the tip scratches

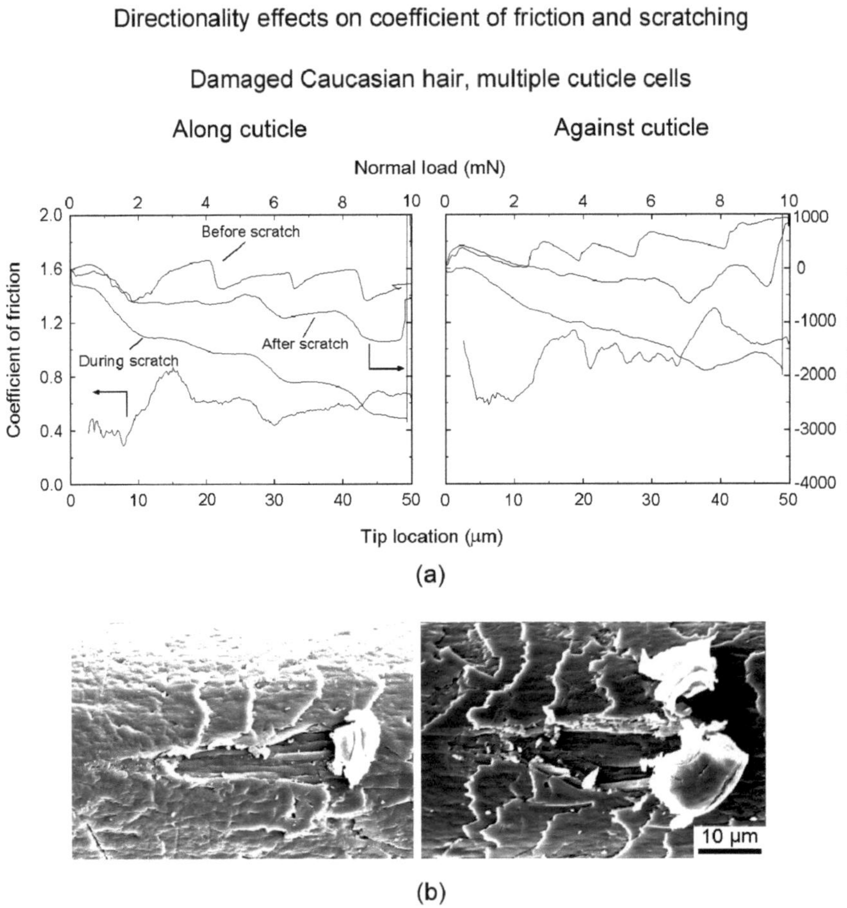

Fig. 4.10 (a) Coefficient of friction and scratch depth profiles as a function of normal load and tip location on multiple cuticle cells of chemo-mechanically damaged Caucasian hair obtained in two scratch tests: scratch along the cuticle and scratch against the cuticle; (b) SEM images of the hair surface after scratch (Wei and Bhushan, 2006)

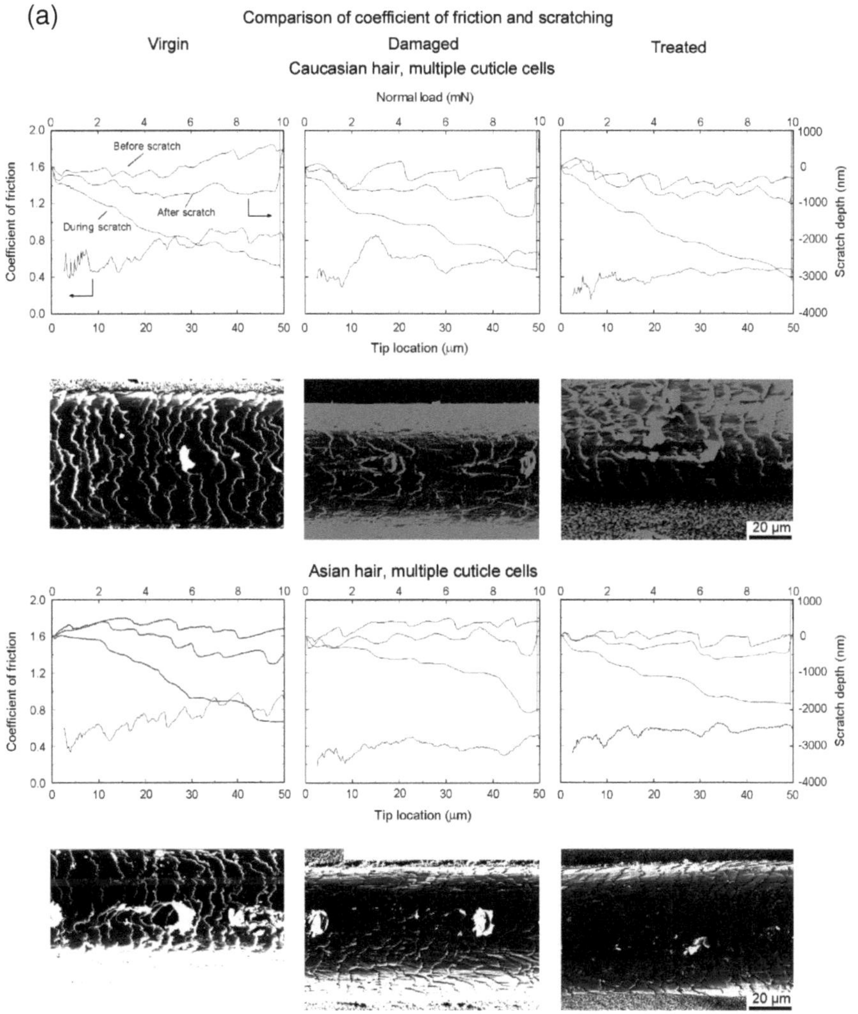

Fig. 4.11 (**a**) Coefficient of friction, scratch depth profiles, and SEM images of Caucasian and Asian hair (virgin, chemo-mechanically damaged, and virgin treated), (**b**) high-magnification SEM images of damaged Caucasian and Asian hair after scratch (Wei and Bhushan, 2006)

the hair surface against the cuticle, the 300–500 nm high cuticle "wall" resists the tip movement, leading to higher coefficient of friction (Wei and Bhushan, 2006).

By observing the surface profiles (before scratch) of Fig. 4.10a, we can clearly see the shape (height and visible length) of each cuticle cell, i.e., the height is about 300–500 nm, and the visible length is about 5–10 μm, which is in good agreement with SEM and AFM data. The scratch tip acts as a surface profiler before scratching. During scratching, the in situ displacement increased up to about 3 μm at 10 mN, while the residual depth is about 1.5 μm. Considering that the thickness of the

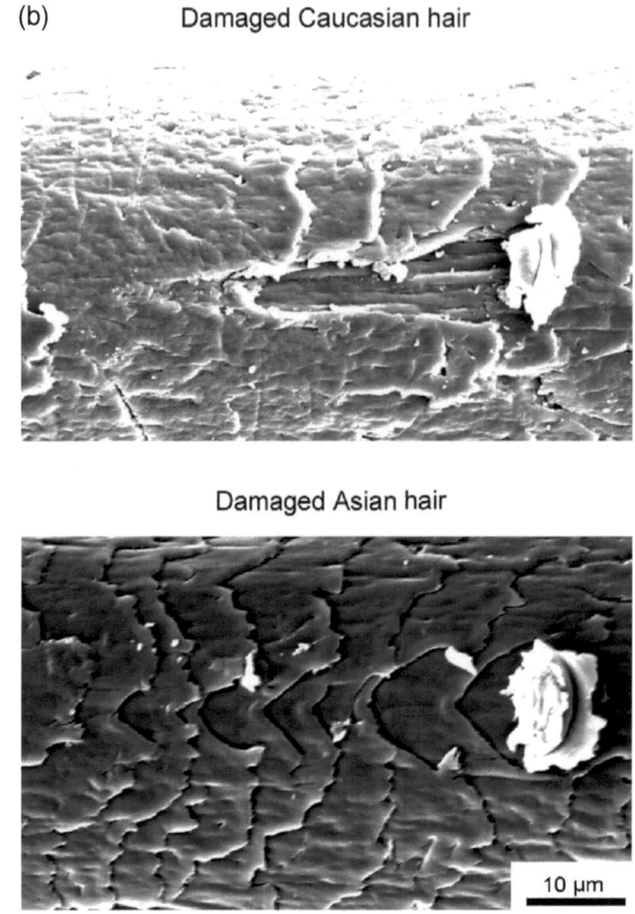

Fig. 4.11 (continued)

cuticle is about 1.5–5 μm, it is likely that during the 10 mN scratch test, the tip reached the cortex. The SEM images (Fig. 4.10b) clearly show that in both the along-cuticle and against-cuticle cases, the cuticle cells were worn away. The topography of the exposed surface is totally different from the cuticle topography, and it is believed that the exposed surface is the cortex. It can also be seen from Fig. 4.10b that the scratching against the cuticle caused much more damage to the hair than along the cuticle.

Given the fact that the "directionality effect" is universal for each human hair from all races and that the scratching along the cuticle is more relevant to our daily life, we now focus on the scratch tests along the cuticle. Figure 4.11a shows the coefficient of friction and scratch depth profiles as a function of normal load and tip location on multiple cuticle cells of Caucasian and Asian hair (virgin, chemo-mechanically damaged, and virgin treated) and the SEM images of the scratch wear tracks. Note that since at least five scratches were made on the same hair, some

SEM images may show more than one scratch wear track. For example, the SEM image of chemo-mechanically damaged Caucasian hair shows two scratches, and the scratch wear track on the right side corresponds to the scratch depth profile. For Caucasian hair, the averaged coefficient of friction of virgin treated hair (~0.4) is lower than virgin hair (~0.7). For Asian hair, the averaged coefficient of friction of virgin treated hair (~0.5) is also lower than virgin hair (~0.8). The trend corresponds well with the nanoscratch results on a single cuticle. Based on this data, it is clear that the conditioner treatment indeed can reduce the coefficient of friction of hair surface upon scratching. Regarding the chemo-mechanically damaged hair, since the coefficient of friction of chemo-mechanically damaged hair varies depending on the type and extent of damage (as discussed above), it is difficult to make a comparison with the virgin or virgin treated hair.

It is worth mentioning that the coefficient of friction of human hair measured by nanoscratch technique is on the microscale and not on the nanoscale, since the tip radius is 1 μm and the normal load range is 1–10 mN. It will be shown later that the coefficient of friction of conditioner-treated hair measured using an AFM tip (radius 30–50 nm) is higher than virgin hair (LaTorre and Bhushan, 2005a). In the nanoscale, the increase in friction force is due in part to an increase in meniscus effects, which increase the adhesive force contribution to friction at sites where the conditioner is deposited or accumulated on the hair surface. This adhesive force is of the same magnitude as the normal load, which makes the adhesive force contribution to friction rather significant. On the microscale, however, the adhesive force is much lower in magnitude than the applied normal load, so the adhesive force contribution to friction is negligible over the hair surface. On the microscale, the conditioner acts as a thin layer of lubricant, decreasing the friction.

Figure 4.11b shows the high-magnification SEM images of chemo-mechanically damaged Caucasian hair and chemo-mechanically damaged Asian hair after scratch. It is interesting to find that the failure mechanisms for these two hair are different. For Caucasian hair, it seems that the tip plowed the cuticle cells continuously during scratching, and the plowed cuticle cells were accumulated at the end of the scratch. As discussed before (also see Fig. 4.10), the exposed surface of the Caucasian hair is believed to be the cortex. For Asian hair, the tip did not plow the cuticle cells continuously. Instead, the tip only broke the top cell of each cuticle and carried away the broken cuticle cells until the end of the scratch. In this case, the tip did not reach the cortex during scratching. In order to explain this, we need to look at the nanomechanical properties of Caucasian and Asian hair. According to mechanical property data reported earlier (Wei et al., 2005), the cuticle of Asian hair has a higher hardness (0.39 ± 0.06 GPa) than Caucasian hair (0.24 ± 0.05 GPa). So the Asian hair may be more "brittle" than Caucasian hair during scratch. That may be the reason why Asian hair is fractured easier than Caucasian hair during scratch. However, it must be noted that our observation is based on a limited number of samples and experiments. Since human hair varies from one hair to another even in the same race, it is hard to draw a general conclusion on hair failure mechanisms in terms of the hair race. What we can say is that the hair fails differently during scratching, depending on the nanomechanical properties of the cuticle of the hair.

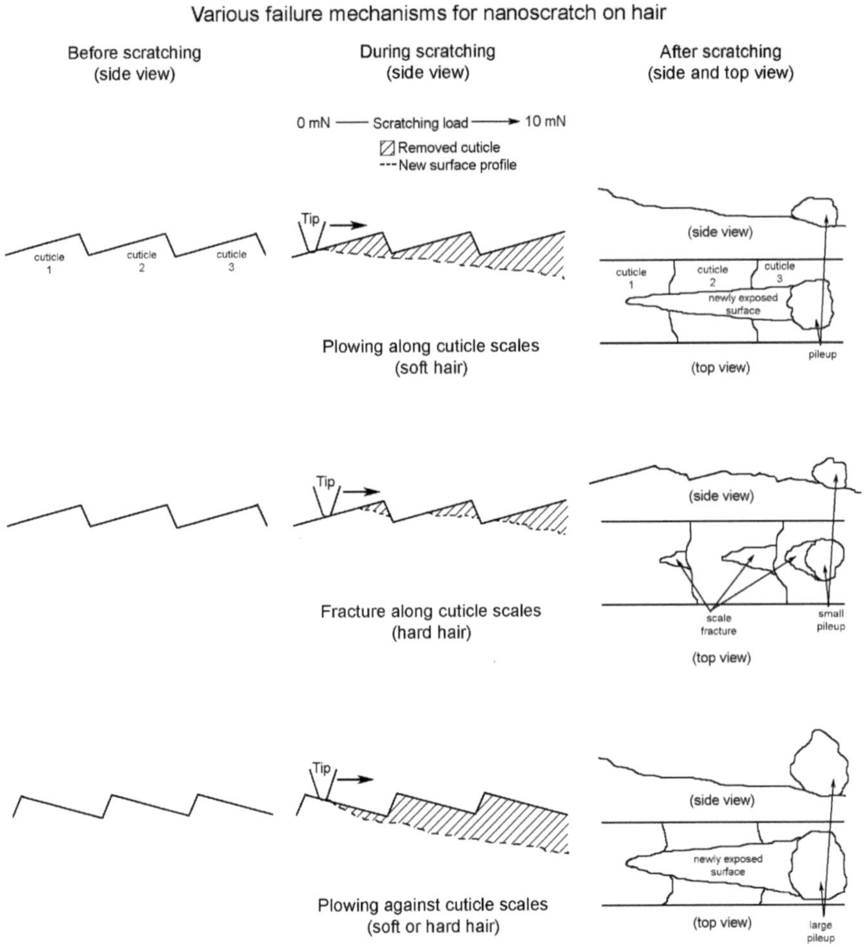

Fig. 4.12 Schematic of the various failure mechanisms during nanoscratching on hair (Wei and Bhushan, 2006)

Figure 4.12 shows the schematic of the various failure mechanisms during nanoscratching on hair. The top and middle figures show the scratch along the cuticle, and the bottom diagram shows the scratch against the cuticle. In the case of scratching along the cuticle, if the hair cuticle is soft (top figure), then the scratch tip will plow the cuticle and carry away the worn cuticle cells (scales). If the load is high enough, then the tip can reach the cortex during scratching. After scratching, a newly exposed surface will be created, and some pileup is formed at the end of the scratch wear track. If the hair cuticle is hard (middle figure), then the scratch tip will fracture the cuticle cells instead of plowing deep into them. After scratching, for each cuticle cell undergoing scratching, part of it is carried away by the tip, resulting in the formation of a small pileup at the end of the scratch wear track and leaving behind a series of incomplete cuticle cells. In the case of scratching against the cuticle (bottom figure), the tip will plow the cuticle cells (whether soft or hard) and create a newly exposed surface with large wedge formation and pileup at the end of the scratch wear track.

In the studies reported earlier, nanomechanical measurements have also been performed on African hair (Wei et al., 2005). In this study, the curly shape and structure made it very difficult to perform the nanoscratch on its surface.

4.2.3 Soaking Effect

Figure 4.13a and b compares the coefficient of friction and scratch depth profiles of unsoaked and soaked Caucasian hair obtained on a single cuticle cell (at 1 mN) and multiple cuticle cells (10 mN loads), respectively. At 1 mN (see Fig. 4.13a), the coefficient of friction of virgin and chemo-mechanically damaged Caucasian hair increased from ~0.4 to ~0.7 after soaking, while the coefficient of friction of virgin treated hair (~0.3) does not change much. It is known that human hair swells in water. In this work, the hair was only soaked in de-ionized water for 5 min. After the sample was soaked, it took a few minutes to mount the sample and run the scratch tests. In this case, it is possible that only a few hundred nanometers of the hair surface contained a considerable amount of water and were softened. During

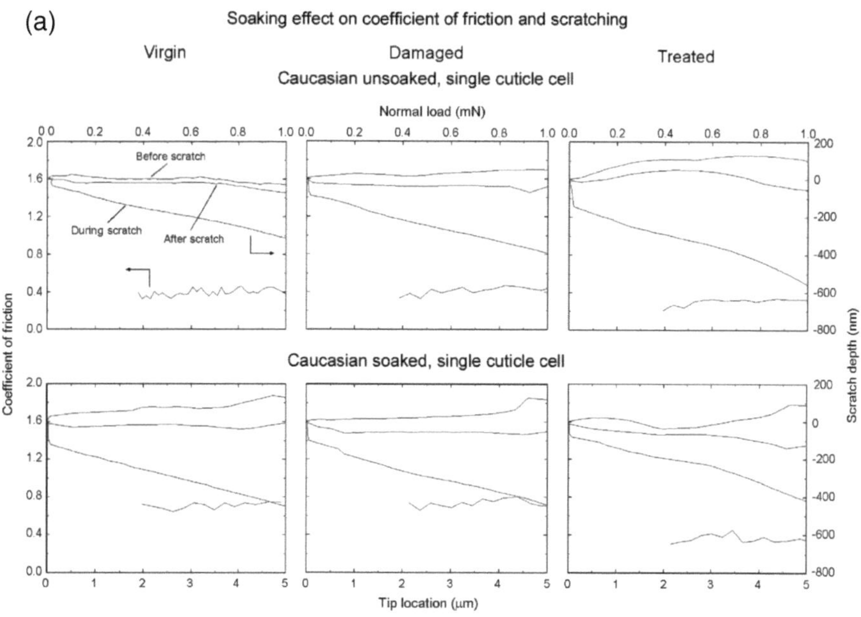

Fig. 4.13 Comparison of nanoscratch results on chemo-mechanically damaged Caucasian unsoaked and soaked hair samples: (**a**) scratch on single cuticle cell (1 mN load); (**b**) scratch on multiple cuticle cells (10 mN load) (Wei and Bhushan, 2006)

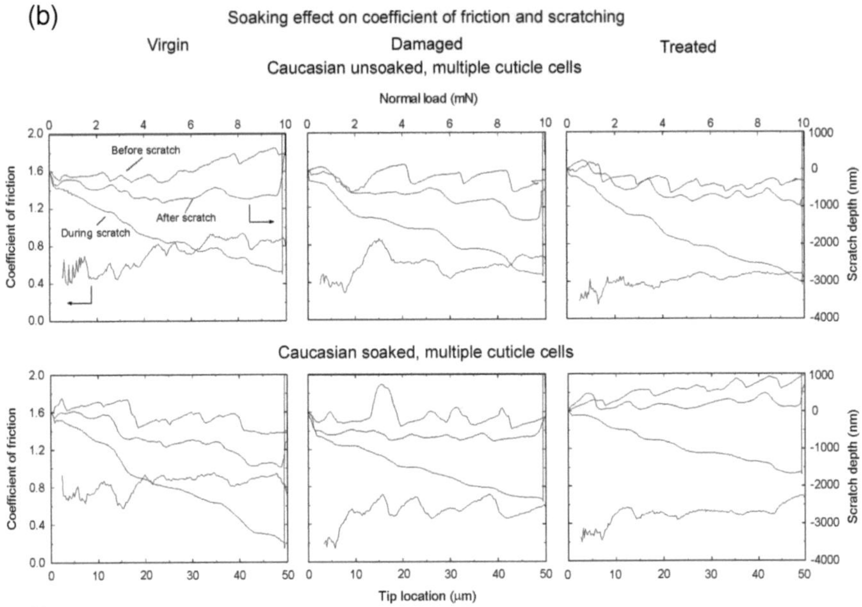

Fig. 4.13 (continued)

scratching, it is easier for the tip to plow into the softer hair surface, leading to higher coefficient of friction. This may be the reason that for 1 mN scratch in which the maximum in situ scratch depths were less than 600 nm, the coefficient of friction of virgin and chemo-mechanically damaged hair increased. For virgin treated hair in the 1 mN scratch test, however, some of the conditioner molecules might occupy the pathways of water molecules so that the swelling of virgin treated hair was not as significant as virgin and chemo-mechanically damaged hair. Therefore, the virgin treated hair shows little change in the coefficient of friction after soaking. At 10 mN (see Fig. 4.13b), the coefficient of friction of all three hair does not change considerably after soaking. This may indicate that the 5 min soaking did not affect the hair surface deeply. Table 4.2 summarizes the coefficient of friction and scratch depths of Caucasian (unsoaked and soaked) and Asian (unsoaked) hair (Wei and Bhushan, 2006).

4.3 In Situ Tensile Deformation Studies on Human Hair Using AFM

In situ tensile deformation experiments are carried out to study the progress of deformation and morphological changes in the hair fiber. Pure tensile and fatigue experiments were conducted (Seshadri and Bhushan, 2008a, b).

Table 4.2 Summary of coefficient of friction and scratch depths of Caucasian (unsoaked and soaked) and Asian (unsoaked) hair

	Unsoaked Caucasian						Soaked Caucasian						Unsoaked Asian					
Max. normal load/no. of cuticle cell	1 mN/single cuticle cell			10 mN/multiple cuticle cells			1 mN/single cuticle cell			10 mN/multiple cuticle cells			1 mN/single cuticle cell			10 mN/multiple cuticle cells		
Hair condition	Virgin	Damaged	Treated	Virgin	Damaged	Treated	Virgin	Damaged	Treated	Virgin	Damaged	Treated	Virgin	Damaged	Treated	Virgin	Damaged	Treated
Average coefficient of friction	0.4	0.4	0.3	0.7	0.6	0.4	0.7	0.7	0.3	0.8	0.6	0.5	0.5	0.6	0.3	0.8	0.4	0.5
Max. in situ depth (nm)	280	440	650	3250	2500	2750	570	540	500	3000	2100	2600	480	500	400	2500	2400	1800
Max. residual depth (nm)	25	100	150	1100	1000	750	130	170	210	1000	500	800	60	150	130	700	900	480

4.3.1 Tensile Deformation of Caucasian Virgin, Damaged, and Treated Hair

Figure 4.14 presents stress–strain curves for five types of hair (Seshadri and Bhushan, 2008a). The stress–strain curve of human hair is similar to that of wool and other such keratinous fibers (Zviak, 1986). When a keratin fiber is stretched, the load elongation curve shows three distinct regions, as marked in Fig. 4.14 (Beyak et al., 1969; Feughelman, 1982). In the pre-yield region, also referred to as the Hookean region, stress and strain are proportional, and an elastic modulus can be found. In this region, there is the homogenous response of α-keratin to stretching. The resistance is provided by hydrogen bonds that are present between turns and stabilize the α-helix of keratin. The yield region represents the transition of keratin from the α-form to the β-form, the chains unfold without any resistance, and hence the stress does not vary with strain. The β-configuration again resists stretching. So, in the post-yield region, the stress again increases with strain until the fiber breaks. This α- to β-transition of keratin is the reason for the unique shape of the stress–strain curve of hair. Typically, yield begins around 5% and post-yield begins at around 15% strain. There is no apparent difference in tensile properties of the different hair types. In the dry state, the tensile properties of hair are significantly contributed by the cortex. Hair fibers oxidized with diperisophthalic acid, which causes almost total removal of the cuticle, have shown no significant change in mechanical properties (Robbins and Crawford, 1991). In the case of conditioner

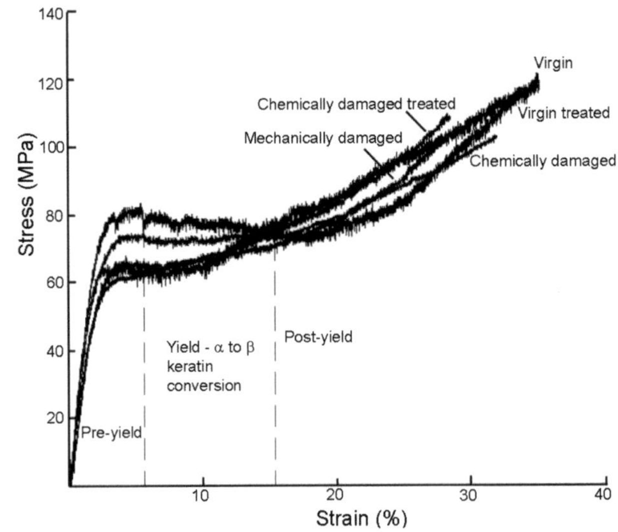

Fig. 4.14 Stress–strain curves for five types of hair – virgin, virgin treated, chemically damaged, chemically damaged treated, mechanically damaged. Transition of α-keratin to β-keratin in the yield region is the reason for the unique shape of the curves (Seshadri and Bhushan, 2008a)

treatment and mechanical damage, the change occurs only in the cuticle. Hence, it is logical that the mechanical properties do not experience a measurable change. In the case of chemical damage, apart from cuticle damage, oxidation of the cystine in keratin to cysteic acid residues occurs, and hence disrupts the disulfide cross-links. However, the disulfide bonding does not influence tensile properties of keratin fibers in the dry state, though some effect is seen in the wet state (Robbins, 1994). Since all experiments in the present study were carried out in the dry state, it is again logical that no change in the mechanical properties was observed.

Figure 4.15 shows AFM topographical images and 2D profiles at indicated planes of a given control area with increasing strain of the virgin, chemically damaged, and mechanically damaged types of hair (Seshadri and Bhushan, 2008a). The common effect of stretching, observed for all hair types, is the lifting of the outer cuticle layer with increasing strain.

In the case of virgin hair (Fig. 4.15a), this is seen to be the only effect of tension. The cuticle lift-off is sudden and occurs consistently at around 20% strain. As mentioned earlier, human hair cuticles have a lamellar structure. The various layers of human hair vary in their cystine content. This causes variation in their mechanical strength. The epicuticle and exocuticle are high in cystine content and are extremely rigid. The endocuticle and cell membrane complex are low in cystine content and are flexible and can extend more. Stretching hair sets up interlayer shear forces due to this difference in extensibility (Reutsch and Weigmann, 1996). At 20% strain, delamination occurs, and the inner cuticle layers separate from the outer ones. This causes outer cuticle lift-off and hence the change in height and slope observed in the AFM images and corresponding cross-sectional profiles.

Figure 4.15b and c shows topographical images of chemically damaged hair and mechanically damaged hair, respectively. Apart from the lift-off discussed earlier, another noticeable effect in both cases is the disappearance of some cuticle edges and the appearance of sponge-like scales instead. Sponge-like scales in the cuticle have been observed earlier. These are thought to be the remains of cuticle cells, which have split up to the endocuticle (Swift and Bews, 1974). Chemical damage and mechanical damage cause cuticle damage and weakening. Hence, in the case of damaged hair, along with failure and delaminations in the endocuticular region, fracture of the outer cuticle occurs to expose the inner cuticle. This shows up as the sponge-like scales observed.

Comparing mechanical damage and chemical damage, it is seen that the fracture occurs earlier in mechanically damaged hair (10%) than in chemically damaged hair (20%). This could be because, in mechanically damaged hair, parts of the cuticle have already broken away, and cuticle damage is more extensive than in chemically damaged hair.

Conditioner in the case of virgin and chemically damaged treated hair has no apparent effect on the lift-off or fracture phenomenon (data not shown). It coats the cuticle and improves its tribological properties. However, it does not chemically or physically alter the cuticles appreciably.

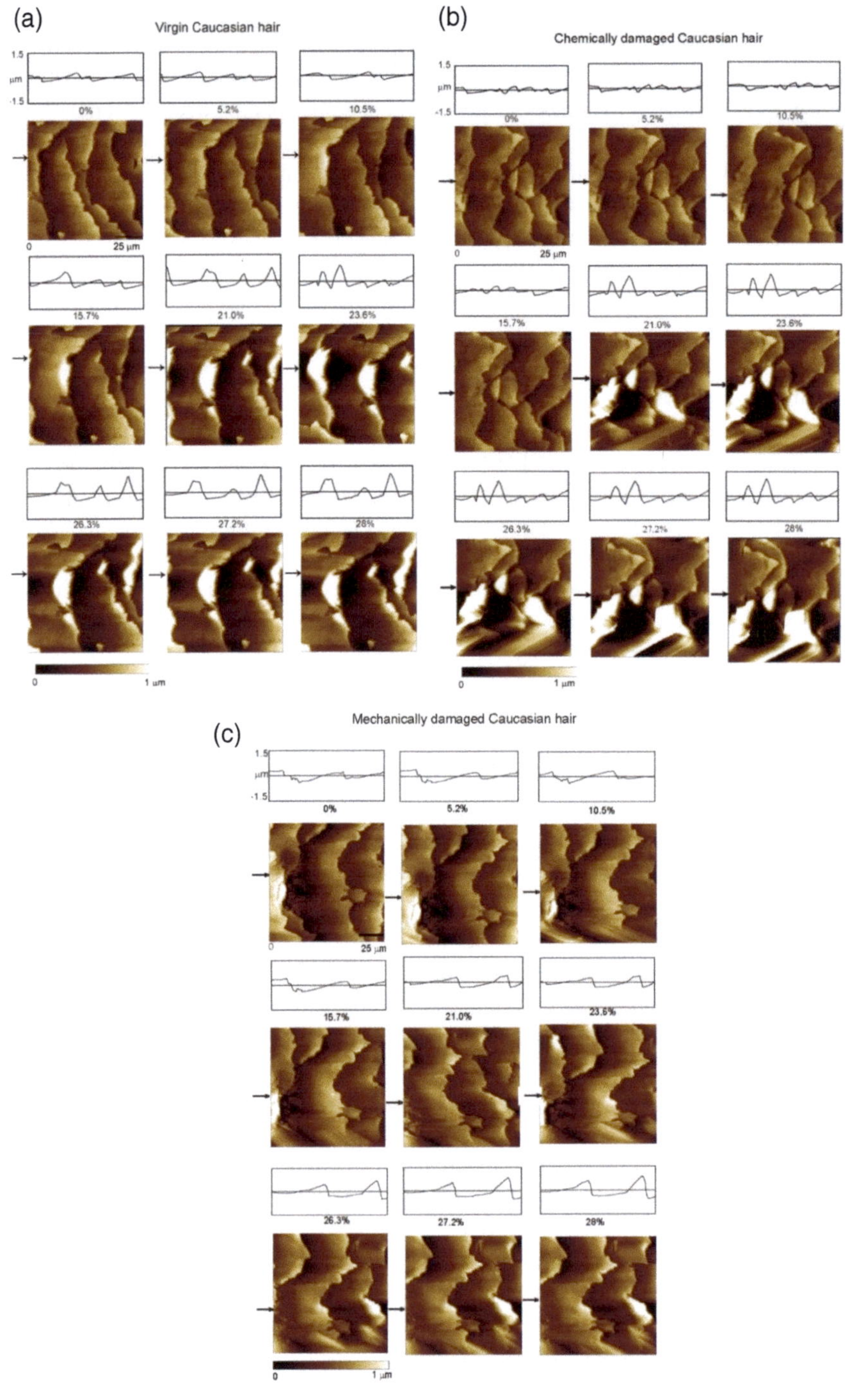
(a)
Virgin Caucasian hair
1.5
µm
-1.5
0%
5.2%
10.5%
0
25 µm
15.7%
21.0%
23.6%
26.3%
27.2%
28%
0
1 µm
(b)
Chemically damaged Caucasian hair
1.5
µm
-1.5
0%
5.2%
10.5%
0
25 µm
15.7%
21.0%
23.6%
26.3%
27.2%
28%
0
1 µm
(c)
Mechanically damaged Caucasian hair
1.5
µm
-1.5
0%
5.2%
10.5%
0
25 µm
15.7%
21.0%
23.6%
26.3%
27.2%
28%
0
1 µm

4.3.2 Effect of Ethnicity on Tensile Deformation

Figure 4.16 presents stress–strain curves for the three categories of ethnic hair – Caucasian, Asian, and African; and Table 4.3 displays the calculated mechanical properties. The elastic modulus is calculated by fitting a line to the linear portion of the curve and finding its slope. The yield stress is taken as the highest point of this linear portion. The strain and stress at the point where the fiber fails are taken as the strain to break and the breaking strength, respectively. As is the case in Fig. 4.14, the stress–strain curve of human hair shows three distinct regions. Typically, yield begins around 5%, and post-yield begins at around 15% strain. The unique shape of the stress–strain curve of human hair is caused by the α- to β-keratin transition.

From the stress–strain curves, it is seen that in general, Asian hair has the highest ultimate strength and strain to break, followed by Caucasian hair. African hair shows very low mechanical properties in terms of modulus, yield point, and strain to break.

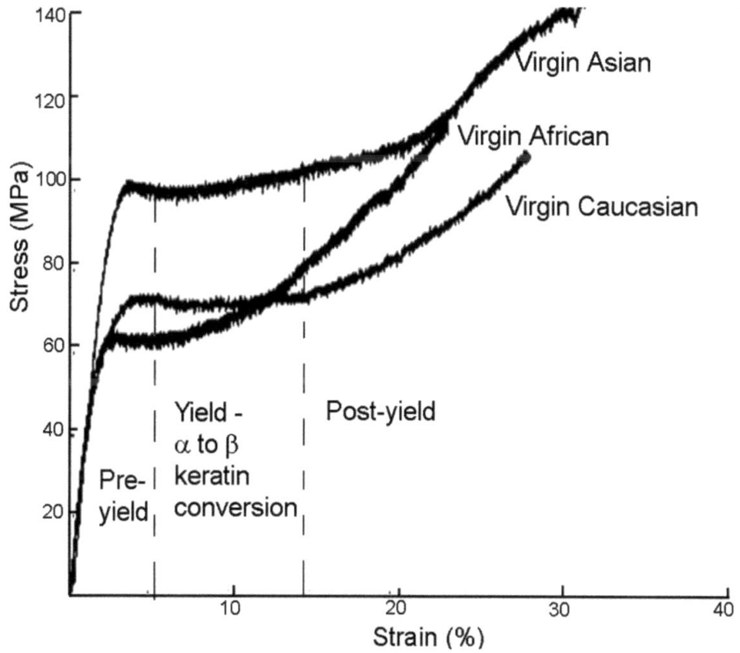

Fig. 4.16 Stress–strain curves for virgin Caucasian, Asian, and African hair. Transition from α- to β-keratin is the reason for the unique shape of these curves

Fig. 4.15 AFM topographical images and 2D profiles at indicated planes of a control area showing progress of damage with increasing strain in (**a**) virgin hair, (**b**) chemically damaged hair, and (**c**) mechanically damaged hair. Cuticle lift-off occurs at about 20% strain in all cases. Fracture of cuticle outer layer occurs at about 20% strain in chemically damaged hair and about 10% strain in mechanically damaged hair (Seshadri and Bhushan, 2008a)

Table 4.3 Mechanical properties of ethnic hair. The values presented here are based on 15 measurements. Values show up to 15% variation from sample to sample

	Caucasian	Asian	African
Elastic modulus (GPa)	3.3	4.7	2.5
Yield strength (MPa)	67	100	58
Breaking strength (MPa)	117	139	101
Strain at break (%)	35	32	20

The lowered mechanical properties of African hair are thought to be partly arising from the higher stresses encountered by it during normal combing and detangling due to its characteristic curly structure (Kamath and Hornby, 1984).

Figure 4.17 shows AFM topographical images and 2D profiles at indicated planes of a given control area with increasing strain of virgin Caucasian, Asian, and African hair. The effect of tension on Caucasian and Asian hair is quite similar. The most significant effect of tension in both cases is lifting of the outer cuticle layer. For virgin Caucasian hair, this inner cuticle failure and outer cuticle lift-off occurs at about 20% strain, whereas for Asian hair noticeable lift-off occurs later at around 24% strain. Also, compared to Caucasian hair, the number of cuticle edges that lift-off is fewer in Asian hair. These differences could be caused because the inner cuticle of Asian hair is also quite rigid. This reduces the extensibility difference between the outer and inner cuticle, and hence the magnitude of the interlayer shear force is reduced.

African hair cuticle shows a complete rupture at around 20% strain. Based on the image at 20% strain, individual cuticle cells are not decipherable. Generally, even virgin African hair cuticles are seen to have undergone a considerable amount of cuticle damage due to combing and detangling. This might cause the cuticle to rupture at ~20%. It is interesting to note that in line with previous observations, most failures of African hair were seen to occur in the regions of twist reversals, as stress concentrations occur in these regions (Kamath and Hornby, 1984).

4.3.3 Effect of Soaking on Tensile Deformation of Caucasian Virgin, Damaged, and Treated Hair

Wet tensile properties of human hair are generally quite different from dry tensile properties. Figure 4.18 shows stress–strain curves for virgin, chemically damaged, and chemically damaged treated unsoaked hair, along with curves of soaked hair of the same three types. Table 4.4 displays the calculated mechanical properties. A stress–strain curve for virgin hair soaked and then dried is also presented.

From the stress–strain curves of the unsoaked samples, it is seen that the curves almost coincide, and there is no significant difference between virgin, chemically damaged, and treated Caucasian hair. The stress–strain curves of the soaked samples are considerably different from unsoaked samples. While the strain to break increases, the yield strength, ultimate strength, and elastic modulus are considerably

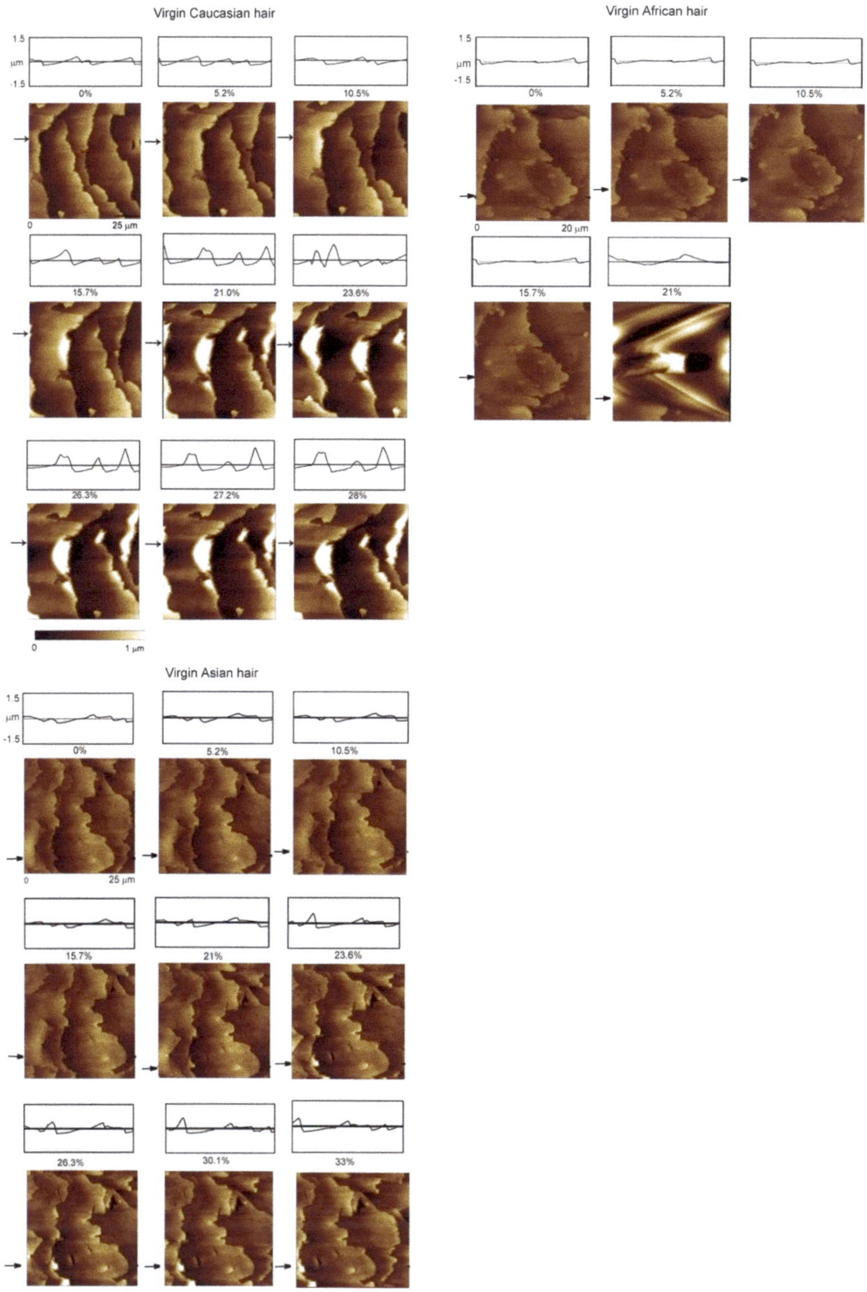

Fig. 4.17 AFM topographical images and 2D profiles at indicated locations of a control area showing progress of damage with increasing strain in Caucasian, Asian, and African hair. Caucasian hair shows cuticle lift-off at ~20% and Asian at ~25%. African hair shows cuticle rupture (Seshadri and Bhushan, 2008b)

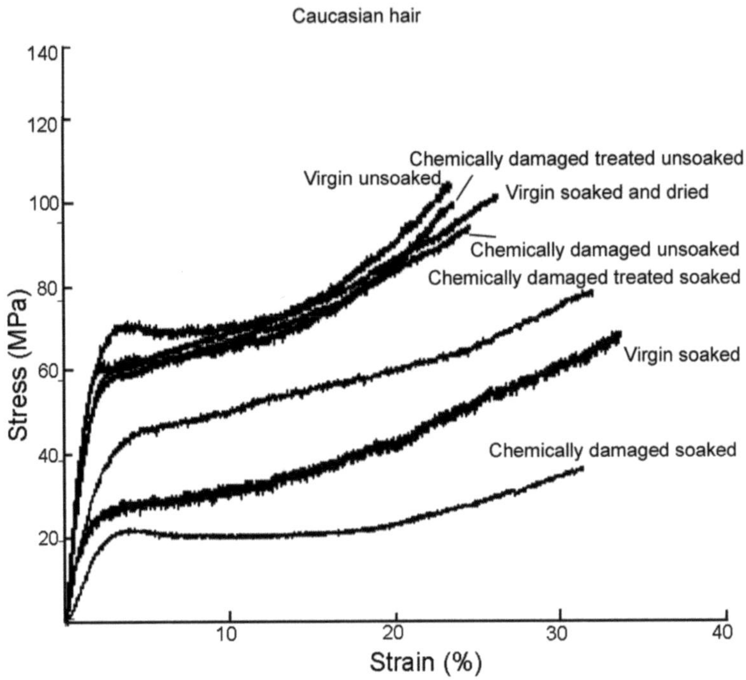

Fig. 4.18 Stress–strain curves for unsoaked and soaked samples of Caucasian virgin, chemically damaged, and chemically damaged treated hair, and virgin soaked and dried hair (Seshadri and Bhushan, 2008b)

reduced by soaking. Soaking apparently softens the hair fiber. One reason for the difference in the mechanical properties is the diametrical swelling of hair in water. Human hair exhibits 14–16% diametrical swelling in water (Robbins, 1994). In the present experiments, the diameter of every fiber (soaked and dry) was measured prior to tension tests as described earlier. Though a comparison between wet and

Table 4.4 Mechanical properties of unsoaked and soaked Caucasian hair. All values are based on 15 measurements. Data show a 15% variation from sample to sample

	Unsoaked			Soaked			Soaked and dried
	Virgin	Chemically damaged	Chemically damaged, treated	Virgin	Chemically damaged	Chemically damaged, treated	Virgin
Elastic modulus (GPa)	3.3	3.3	2.9	0.9	0.5	0.9	3.1
Yield strength (MPa)	67	61	74	28	23	47	65
Breaking strength (MPa)	117	107	108	70	37	82	106
Strain at break (%)	35	32	31	41	37	39	32

dry diameter of any particular fiber was not carried out, the diameter of the wet fiber was as such used for stress computation in soaked curves, and the dry diameter for unsoaked curves. This leads to the lowered stress values on the stress–strain curves. However, apart from the geometric effect of the diameter, chemical changes are also thought to be responsible for this difference. It has been reported that stress–strain curves of soaked and unsoaked fibers, where stress is computed using only the wet diameter for both fibers, still show considerable difference in mechanical properties and lowering of elastic modulus (Feughelman, 1997). Using the same diameter for both samples is justified as a form of normalization. The basis is that there is still the same amount of structural material in both soaked and unsoaked states. This finding is an indication that the elastic modulus and mechanical properties of soaked fibers are also contributed by chemical interactions. Feughelman (1997) postulates that, while the elastic modulus in the linear portion is mainly due to resistance of the α-helix hydrogen bonds, Coulombic interaction bonds in the α-helix also have a degree of contribution. In the wet state, some of the groups involved in the Coulombic bonding interact with water instead and are broken, which leads to a lowered wet elastic modulus. The changes in mechanical properties in the wet state have also been attributed to the presence of high sulfur and high glycine tyrosine protein groups (Zviak, 1986). In the study reported here, the geometric effect due to usage of different diameters for the wet and dry state and the chemical changes in the fiber itself may both be responsible for the differences in the mechanical properties observed in the stress–strain data. Among the soaked samples, chemically damaged hair has the lowest mechanical properties, while treated hair has slightly better mechanical properties. Chemically damaged hair is more hydrophilic than virgin hair. Conditioner coats the surface of damaged hair and makes the surface less hydrophilic (LaTorre and Bhushan, 2006). Hence, chemical damage renders the cuticle more permeable to water, leading to higher diametrical swelling. In chemically damaged treated hair, it is also possible that conditioner molecules may be occupying the path of water molecules (Wei et al., 2005), leading to lesser diametrical swelling. Virgin hair which has been soaked and dried shows a full recovery of mechanical properties. This again could be because the Coulombic bonding is restored when water is removed, and also the diameter decreases.

Figure 4.19 shows a schematic of the effect of soaking on hair morphology of virgin, chemically damaged, and chemically damaged treated Caucasian hair.

4.3.4 Fatigue Studies of Caucasian Virgin, Damaged, and Treated Hair

Figure 4.20 presents stress–strain curves for fatigued virgin, damaged, and treated Caucasian hair fibers, and Table 4.5 presents the calculated mechanical properties. It is seen that fatigue softens the hair fiber and reduces its mechanical properties and the strain to break point. A possible explanation could be that low load fatigue reduces intercellular adhesion in the cortical cells (Kamath and Hornby, 1984). This

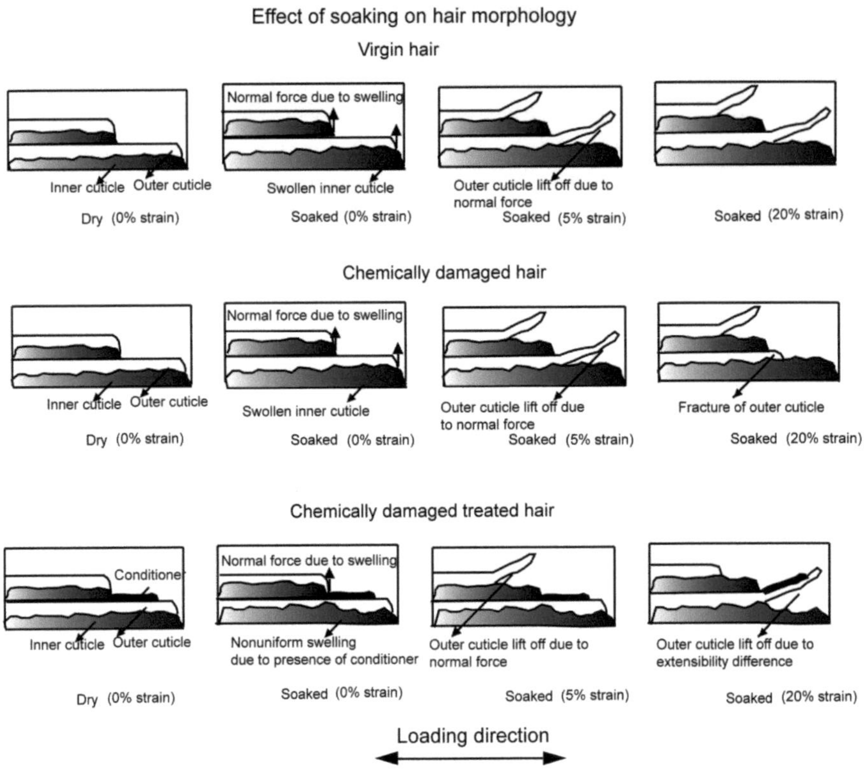

Fig. 4.19 Schematic diagram of the effect of soaking on hair morphology (Seshadri and Bhushan, 2008b)

weakening of the cortex leads to the overall reduction in mechanical properties. Again, no significant difference is seen between virgin, damaged, and treated hair. The curves almost coincide. The reasons for this are most probably the same as that for unsoaked virgin, damaged, and treated hair – the tensile properties in the dry state are significantly determined by the cortex, and chemical damage and conditioner treatment predominantly affect the cuticle. Hence no apparent difference is seen in the stress–strain curves.

It is interesting to note that African hair shows very similar behavior to damaged fatigued hair. Figure 4.20 shows the stress–strain curve of African hair along with the fatigue samples. Full cuticle rupture is seen in African hair. Also it shows a strain to break point comparable to fatigued hair. The shape of the stress–strain curve is also similar to fatigued hair, except that the mechanical properties are higher. These similarities could be because African hair undergoes considerable cuticle damage and fatigue due to its tendency to tangle, even during normal grooming processes. This causes it to show behavior similar to fatigued and damaged Caucasian hair. The

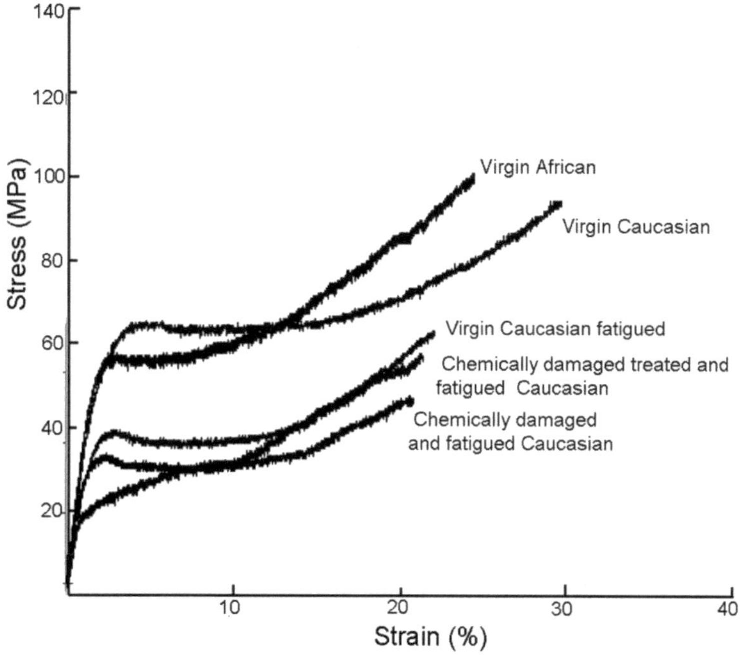

Fig. 4.20 Stress–strain curves for virgin, chemically damaged, and chemically damaged treated Caucasian hair subjected to fatigue. Virgin Caucasian and virgin African stress–strain curves are shown for comparison (Seshadri and Bhushan, 2008b)

mechanical properties of African are probably higher because hair recovers some of its mechanical properties on relaxation.

It has also been shown in this body of work that no visible change occurs in the surface morphology in 5000 cycles of straining, with a maximum strain of ~15%. This could be because the first indication of failure in the dry state, cuticle lift-off, occurs at ~20% strain, and the maximum cyclic strain is kept below this value.

Table 4.6 presents a summary of observations and mechanisms related to the response of ethnic (Caucasian, Asian, and African) and soaked and fatigued Caucasian hair fibers subjected to tensile stress.

Table 4.5 Mechanical properties of fatigued Caucasian hair. The values presented here are based on five measurements. Values show up to 15% variation from sample to sample

	Virgin	Chemically damaged	Chemically damaged treated
Elastic modulus (GPa)	1.6	1.1	1.5
Yield strength (MPa)	36	24	42
Breaking strength (MPa)	63	52	70
Strain at break (%)	20	20	21

Table 4.6 Summary of failure modes of various hair samples

Test matrix		Strain at break (%)	Observations
Ethnicity	Caucasian	31	Appreciable outer cuticle lift-off occurs at ~20% strain due to outer cuticle failure No cuticle rupture/fracture occurs
	Asian	35	Appreciable outer cuticle lift-off occurs at ~25% strain due to inner cuticle failure No cuticle rupture/fracture Lift-off area is lesser than Caucasian hair
	African	20	Lowest strain to failure Full rupture of cuticle and fiber failure occurs at ~20% strain Failure mode is similar to fatigued and chemically damaged Caucasian hair
Soaking (Caucasian)	Virgin	38	Cortex and inner cuticle swell appreciably and hair is softened Outer cuticle lift-off occurs at ~5% strain due to swollen inner cuticle No cuticle rupture/fracture
	Virgin soaked and dried	31	All properties are same as virgin Caucasian hair Full recovery on drying
	Chemically damaged	36	Cortex and inner cuticle swell appreciably and hair is softened Outer cuticle lifts off at ~5% strain due to swollen endocuticle Outer cuticle fractures at ~20% strain
	Chemically damaged and treated	35	Non-uniform swelling due to presence of conditioner Some of the outer cuticle edges lift off at ~5% strain and additional ones at ~20% strain Outer cuticle fractures at ~20% strain
Fatigue (Caucasian)	Virgin	20	Weakened cuticle and fiber softening occurs Outer cuticle fractures and fiber fails at ~20% strain No cuticle rupture occurs
	Chemically damaged	19	Damaged cuticle is further weakened, fiber softening Outer cuticle ruptures and fails at ~20% strain
	Chemically damaged and treated	20	Damaged cuticle is further weakened, fiber softens Outer cuticle ruptures and fails at ~20% strain

4.4 Summary

A nanoindenter has been used to perform nanomechanical studies on human hair. The chemical damage and conditioner treatment caused the hardness and elastic modulus of the hair surface to decrease within a depth of less than 1.5 μm. That is, the first 3–4 cuticle scales may interact with the chemicals and conditioner ingredients more effectively than the rest of the scales. It is found that the hair cuticle has a higher hardness and elastic modulus than the cortex in the lateral direction. The hardness and elastic modulus of hair decreased as the indentation depth increased. The cystine content variations in cuticle substructures (A-layer, exocuticle, endocuticle, cell membrane complex) and cortex are proposed to be responsible for the observation. Humidity and temperature had an effect on mechanical properties. The Young's modulus of damaged treated hair decreases dramatically at high humidity. Little temperature effect was observed on the Young's modulus.

Nanoscratch tests were performed on single and multiple cuticles of various hair, in both unsoaked and soaked conditions. The coefficient of friction of virgin treated hair is lower than virgin hair for Caucasian and Asian hair in both cases of single cuticle scratch and multiple cuticle scratch. This thin conditioner layer acts as a layer lubricant, reducing the coefficient of friction during scratching. In situ displacement (30–200 nm) increased greatly at a very low initial load and then increased gradually, indicating that the first approximately 200 nm of the hair surface should be softer than the underlying layer. The nanoscratch tests on multiple cuticles clearly show the directionality effect on the coefficient of friction. It is found that the hair surface fails differently during scratching, depending on the nanomechanical properties of the cuticle of the hair. For a hair with a hard cuticle, the cuticle cells tend to be fractured during scratching. For a hair with a soft cuticle, the scratch tip usually plows and wears away the cuticle cells continuously until it reaches the cortex. The effect of 5 min soaking in de-ionized water on the coefficient of friction and scratch resistance of human hair is limited within a shallow region (about 600 nm deep) of the hair surface. In this case, the coefficient of friction of virgin and chemo-mechanically damaged Caucasian hair increases after soaking because of the swelling of the water, which softens the hair surface.

Human hair shows a stress–strain curve typical of keratinous fibers. Transition of α-keratin to β-keratin in the yield region is the reason for the unique shape of the curves. Chemical damage, mechanical damage, and conditioner treatment have no obvious effect on the stress–strain curve or tensile properties. This is because such treatments affect the cuticle predominantly, and tensile properties of human hair in the dry state are governed by the cortex. Tensile stress in general causes lift-off of the outer cuticle. The lift-off is sudden and occurs consistently at around 20% strain. This lift-off occurs in all types of hair studied and is due to interlayer shear forces and consequent separation between inner and outer cuticle layers at 20% strain. Chemical damage and mechanical damage cause weakening of the outer cuticle. Along with lift-off, fracture of the outer cuticle to expose endocuticular layers occurs. Fracture occurs sooner (about 10% strain) in mechanically damaged hair than chemically damaged hair (about 20% strain). Different ethnic hair types show considerably different mechanical properties. Caucasian hair shows outer cuticle lift-off at ~20% and Asian hair at ~25% strain. African hair shows full cuticle rupture and fiber failure at ~20% strain and behavior similar to damaged, fatigued Caucasian hair. Fatigue has been shown to weaken the cuticle and the cortex.

Chapter 5
Multi-scale Tribological Characterization

5.1 Macroscale Tribological Characterization

Tribology is very important to hair care and product development. While the current state of the art is to use an AFM to measure the nanoscale tribological properties of hair in contact with an AFM tip, macroscale tribological measurements provide an excellent simulation of skin–hair and hair–hair contacts (Bhushan et al., 2005). The friction and wear of hair were measured using a flat-on-flat tribometer. Friction and wear studies on various hair are presented, including effect of load, velocity, and skin size. In addition, the effect of humidity and temperature on hair tribological properties is discussed.

5.1.1 Friction and Wear Studies of Various Hair

Figure 5.1a shows the coefficient of friction measured from hair strands sliding against a polyurethane film (simulated skin) (Bhushan et al., 2005). The data show that the coefficient of friction of virgin Caucasian hair was about 0.14 along the cuticle and about 0.23 against the cuticle. As with most animal fibers, human hair shows a directionality friction effect; that is, it is easier to move a surface over hair in a root-to-tip direction than in a tip-to-root direction because of the anisotropic orientation of the hair cuticles (Bhushan, 1999a, 2002; Robbins, 1994). The data show that the flat-on-flat tribometer can measure the directionality dependence of friction. Note that in Fig. 5.1a, hair strands were used, and all the hair was separated from each other, so during the friction test, there was no interaction between hair and hair. The output signals of normal force and friction force are smooth, and the coefficient of friction has a small variation.

In industry, many friction tests of hair are performed on a bundle of hair, in which some hair is overlapping on each other. So the hair–hair interaction occurs during the friction test, and variation in the data is large. Figure 5.1b shows the coefficient of friction obtained from a bundle of hair. It can be seen that the output signal of normal force fluctuated a lot, and the output signal of friction force is not smooth, leading to a big variation of the coefficient of friction. The coefficients of friction along cuticle and against cuticle are greater for a bundle of hair compared

B. Bhushan, *Biophysics of Human Hair*, Biological and Medical Physics, Biomedical Engineering, DOI 10.1007/978-3-642-15901-5_5,

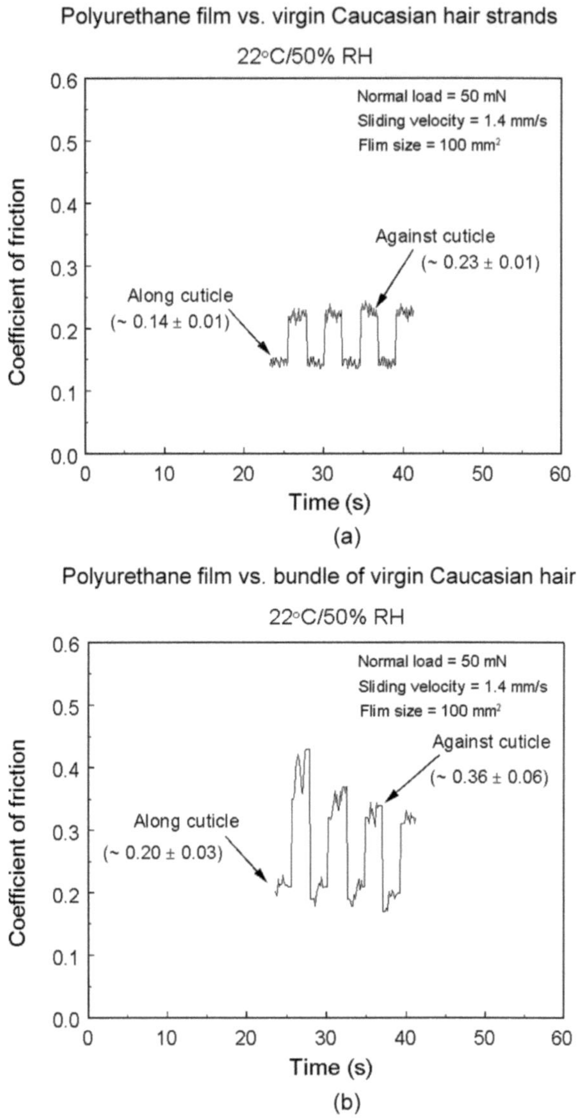

Fig. 5.1 Coefficient of friction measured (**a**) from hair strands and (**b**) from bundle of hair sliding against polyurethane film (Bhushan et al., 2005)

to hair strands, because of the hair–hair interaction during friction tests. It has been observed that if a bundle of hair is used for friction tests, the data have much poorer reproducibility than if hair strands are used. This may be because when a bundle of hair is used, the hair is placed randomly, and the hair to hair position is hard to repeat, but for hair strands, the hair to hair position is easy to control since they are separated and parallel to each other. Therefore, in the chapter, hair strands were used to make further measurements.

Figure 5.2a shows the effect of normal load, sliding velocity, and film area on the coefficient of friction of hair. All coefficient of friction values were obtained

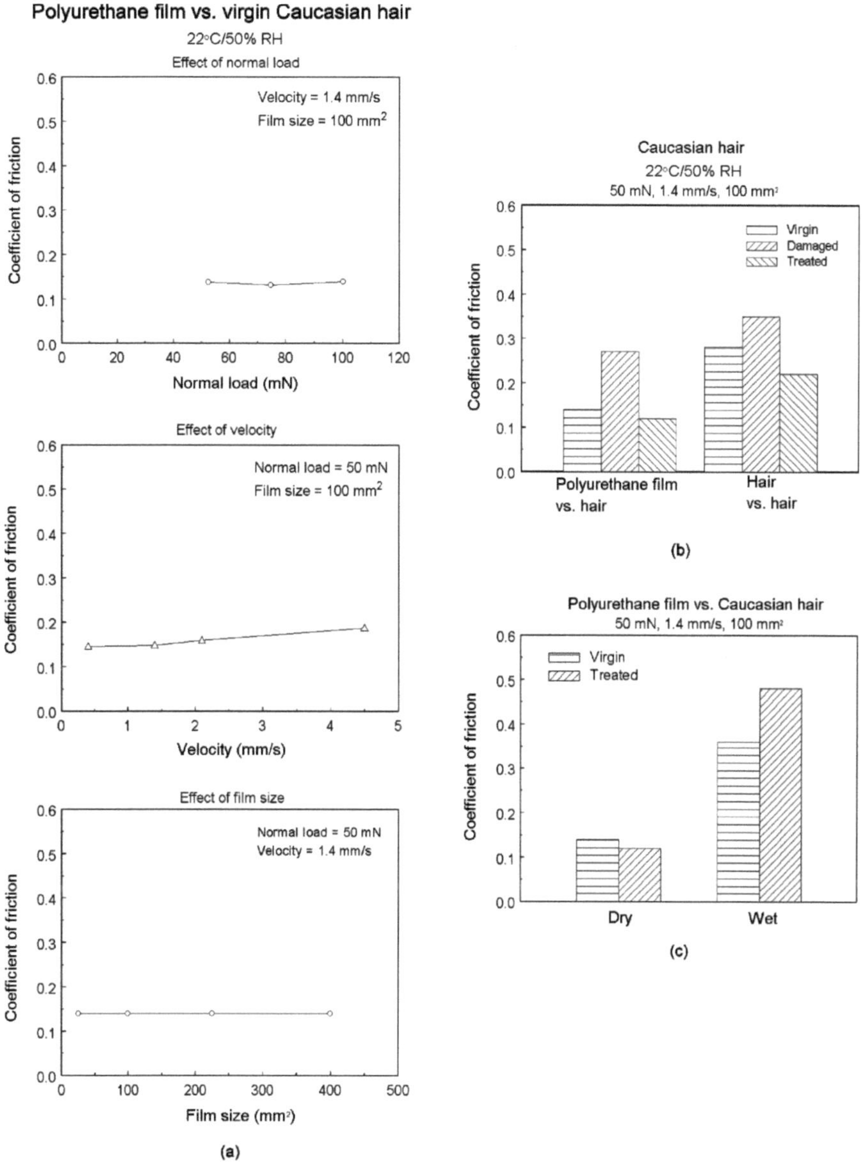

Fig. 5.2 (**a**) Effect of normal load, velocity, and film size on coefficient of friction of polyurethane film vs. virgin Caucasian hair; (**b**) coefficient of friction of polyurethane film vs. Caucasian virgin and treated hair and hair vs. hair; (**c**) coefficient of friction of polyurethane film vs. Caucasian virgin and treated hair at dry and wet conditions (Bhushan et al., 2005)

along the cuticle. It shows that load and film area has no effect on hair friction, but a higher velocity leads to a higher coefficient of friction. According to some observations, coefficient of friction is independent of normal load, apparent area of contact between the contacting bodies, and velocity (Bhushan, 1999a, 2002). Macroscale hair friction appears to obey the first two observations. The third rule of friction, which states that friction is independent of velocity, is not valid in the case of hair friction. Changes in the sliding velocity may result in a change in the shear rate, which can influence the mechanical properties of the hair and polyurethane film (shear strength, elastic modulus, yield strength, and hardness) (Bhushan, 1984). If the mechanical property change can lead to a lower strength of the hair surface and higher real area of contact, then the coefficient of friction will increase.

Figure 5.2b shows the coefficient of friction of polyurethane film vs. hair and hair vs. hair. In the case of polyurethane film vs. hair, the chemo-mechanically damaged hair has the highest coefficient of friction, followed by virgin and virgin treated hair, indicating that the conditioner can reduce the friction of hair. The hair vs. hair results show the same trend as polyurethane film vs. hair, but the coefficient of friction of hair vs. hair is higher than the corresponding polyurethane film vs. hair. Figure 5.2c compares the coefficient of friction of polyurethane film vs. Caucasian hair at dry and wet conditions. Obviously, the coefficient of friction at wet conditions is higher than dry conditions, due to the swelling of hair. When the hair is swollen, the hair cuticle will be lifted up, and the real contact area will be increased, leading to higher coefficient of friction.

Figure 5.3 shows the coefficient of friction of polyurethane vs. various hair. For all hair, the chemo-mechanically damaged hair has the highest coefficient of friction, followed by virgin hair and virgin treated hair. Note that the coefficient of friction can vary about 10–15% within a given ethnic group. Figure 5.4a shows the coefficient of friction of Caucasian hair during wear tests. The coefficient of friction does not change during 24 h for both virgin and virgin treated hair. From the optical micrographs in Fig. 5.4b, it can be seen that after wear tests, some cuticles were damaged. Polyurethane film is soft and does not create much damage to hair. Since the observed damaged area was small, it may not affect the overall coefficient of friction.

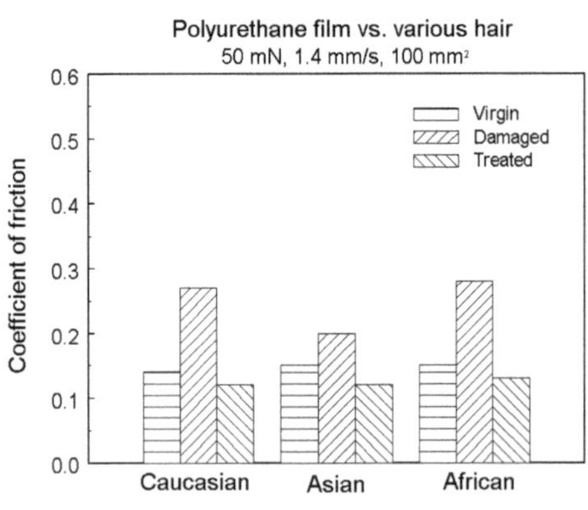

Fig. 5.3 Coefficient of friction of polyurethane film vs. various Caucasian virgin, chemo-mechanically damaged, and virgin treated hair (Bhushan et al., 2005)

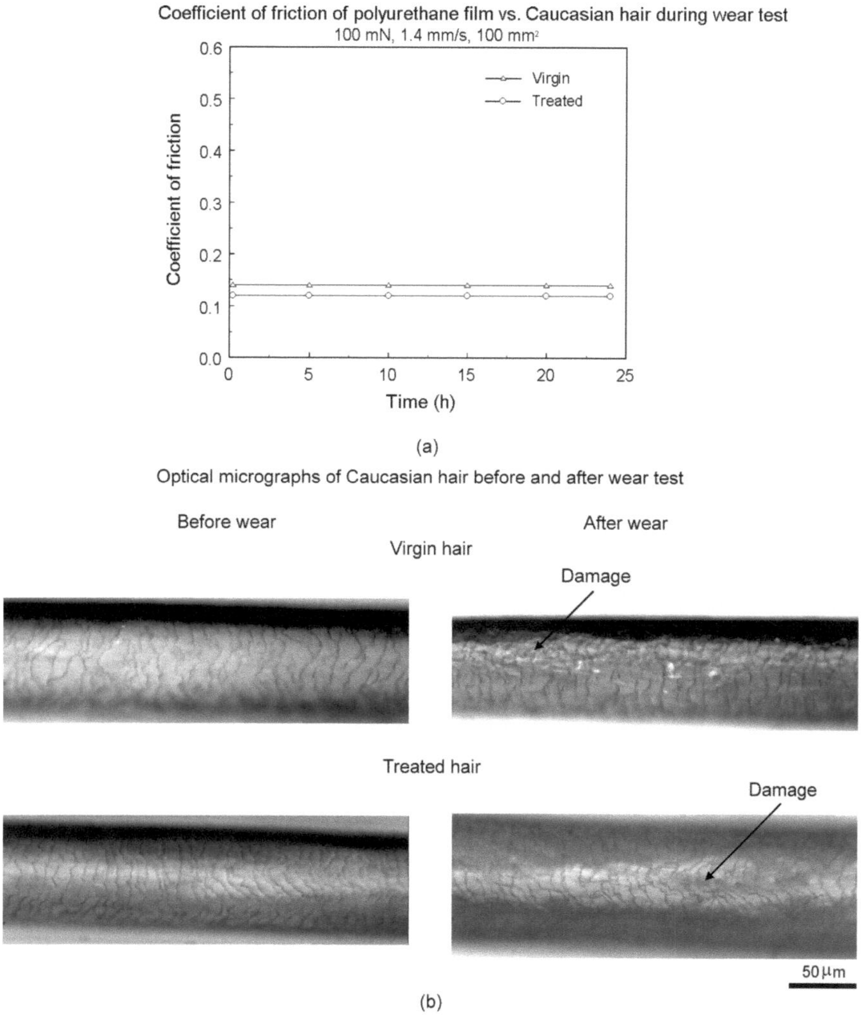

Fig. 5.4 (**a**) Coefficient of friction of polyurethane film vs. virgin and virgin treated Caucasian hair during wear test, (**b**) optical micrographs of virgin and virgin treated Caucasian hair before and after wear test (Bhushan et al., 2005)

5.1.2 Effect of Temperature and Humidity on Hair Friction

Figure 5.5 shows the effect of temperature and humidity on hair friction. The coefficient of friction of hair is a strong function of humidity, and it increases as the relative humidity increases. In addition, the differential friction effect also increases with increasing relative humidity, as shown in Fig. 5.5. It is interesting to find that the differential friction effect of virgin treated hair is less dependent on relative humidity than virgin hair. This may be because some of the conditioner molecules

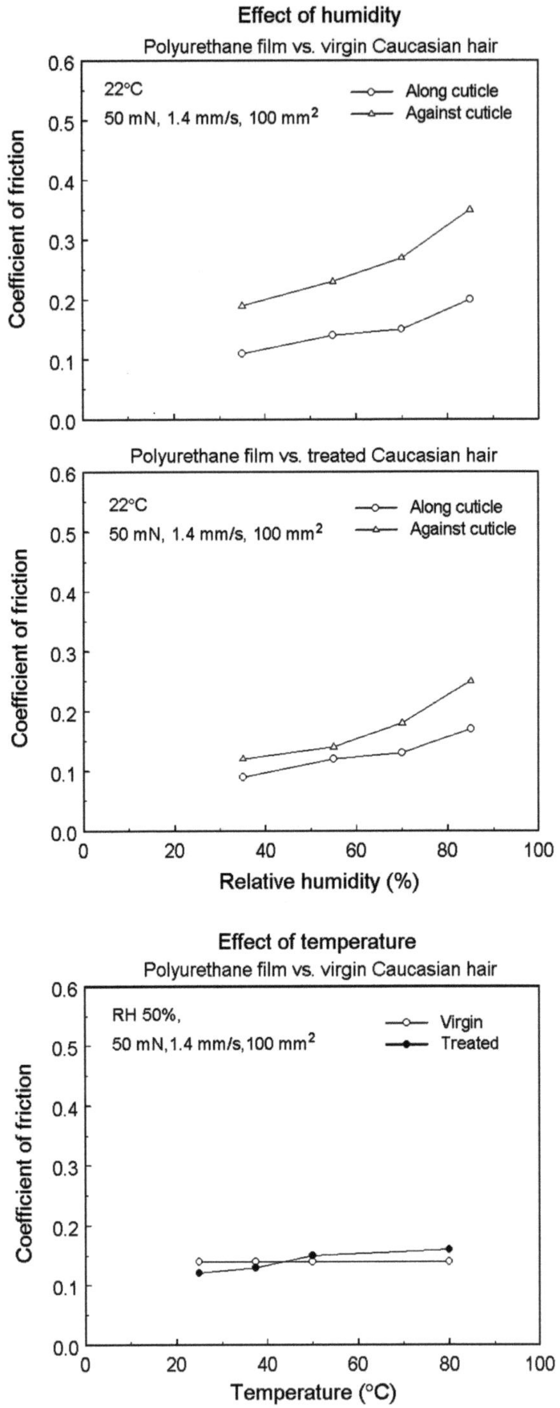

Fig. 5.5 Effect of humidity and temperature on coefficient of friction of polyurethane film vs. virgin and virgin treated Caucasian hair (Bhushan et al., 2005)

occupied the pathways of water molecules so that the swelling of hair was not as significant as virgin hair.

For virgin hair, the temperature has no effect on the coefficient of friction. This is in agreement with Scott and Robbins (1980). For virgin treated hair, it is found that the coefficient of friction increases as the temperature increases. After the hair was treated with conditioner, the hair surface properties might have changed, which could be affected by temperature. For instance, a higher temperature might lead to softer treated hair surface, leading to a higher coefficient of friction.

5.2 Nanotribological Characterization Using an AFM

Nanoscale tribological characterization is essential to study the hair and evaluate/develop better cosmetic products. This becomes especially important when studying the effects of damage and conditioner treatment. How common hair care products, such as conditioner, deposit onto and change hair roughness, friction, and adhesion are of interest, since these properties are closely tied to product performance. Other important issues, such as the thickness distribution of conditioner on the hair surface, which is important in determining the proper functions of conditioner, have been important to cosmetic scientists for decades.

In this section, the roughness, friction, adhesion, and wear of various hair are studied (LaTorre and Bhushan, 2005a, b; LaTorre and Bhushan, 2006; Lodge and Bhushan, 2006a). In Chap. 6, a new method based on the AFM technique for determining thin liquid film thickness is developed, and the conditioner thickness distribution, adhesive force, and the effective Young's modulus of various hair samples are presented (Chen and Bhushan, 2005, 2006; Lodge and Bhushan, 2006a). The binding interactions between the conditioner molecules and hair surface are discussed as well.

5.2.1 Various Ethnicities

Topographical images of Caucasian, Asian, and African hair were taken with scan sizes of $20 \times 20\,\mu m^2$, as shown in Fig. 5.6. Lighter areas of the images correspond to higher topography, and darker areas correspond to lower topography. Only virgin and chemo-mechanically damaged hair are shown in Fig. 5.6 because virgin treated samples closely resemble virgin hair samples. One can see the variation in cuticle structure even in virgin hair. Cracking and miscellaneous damage at the cuticle edges is evident in both virgin and chemo-mechanically damaged conditions. In virgin hair, the damage is likely to be caused by mechanical damage resulting from daily activities such as washing, drying, and combing. Most of the virgin cuticle scales that were observed, however, were relatively intact. Long striations similar to scratches and "scale edge ghosts" (outlines of a former overlying cuticle scale edge left on the underlying scale before it was broken away) were found on the surface.

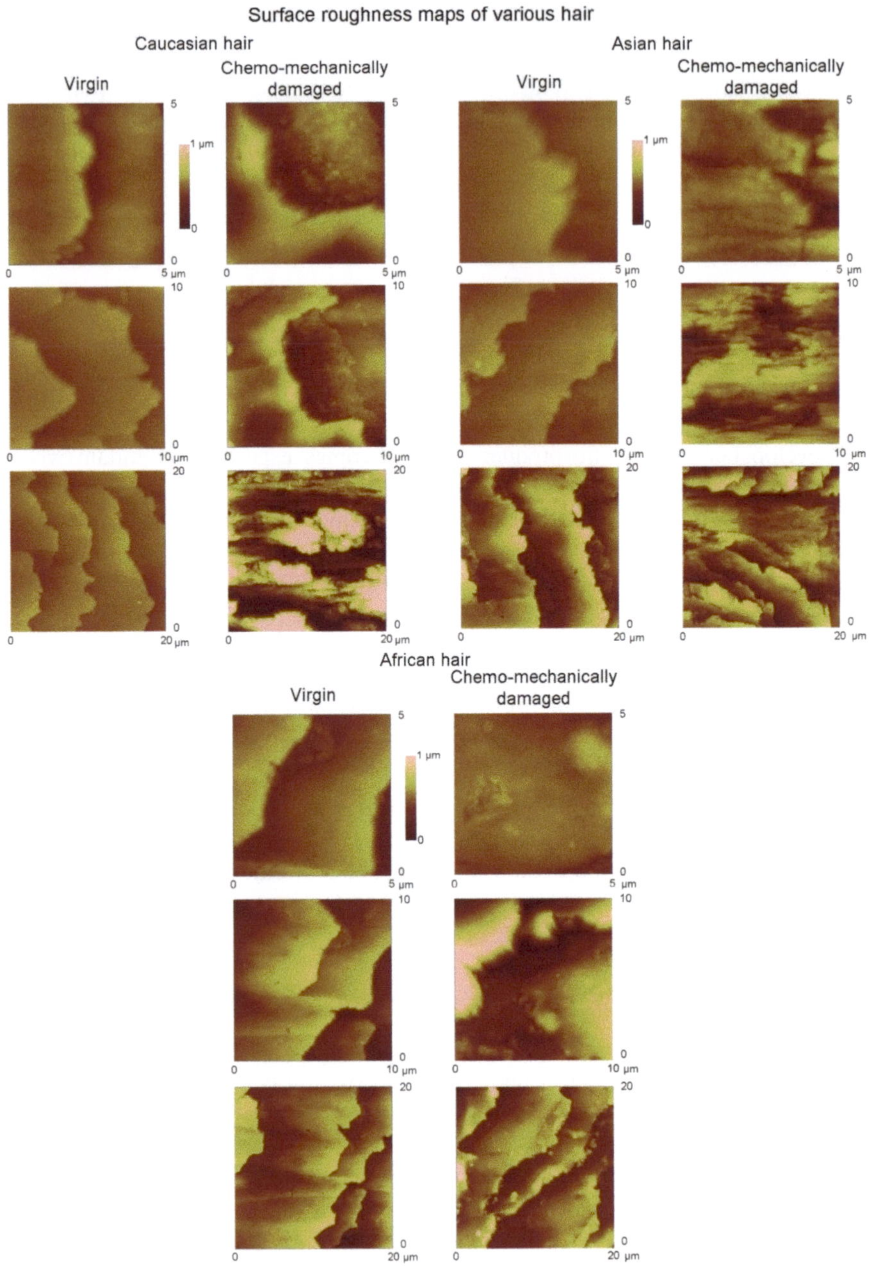

Fig. 5.6 Surface roughness of virgin and chemo-mechanically damaged Caucasian, Asian, and African hair at 5, 10, and 20 μm^2 scan sizes (LaTorre and Bhushan, 2005a)

In some instances, the areas surrounding the cuticle edges appeared to show residue or debris on the surface, which is most likely remnants of a previous cuticle or the underside of the cuticle edges that are now exposed (such as the endocuticle). Caucasian and Asian virgin hair displayed similar surface structure, while the African hair samples showed more signs of endocuticular remains along the scale edges. One can also see more curvature in the cuticle scales of African hair, which is attributed to its elliptical cross-sectional shape and curliness, which can partially uplift the scales in different places. With respect to chemo-mechanically damaged hair, it is observed that several regions seem to exist in these hair samples, ranging from intact cuticle scales to high levels of wear on the surface. In many cases these regions occur side by side. This wide variation in chemo-mechanically damaged cuticle structure results in a wider range of tribological properties on the micro/nanoscale for these fibers. Caucasian and Asian chemo-mechanically damaged hair showed more worn-away cuticle scales than chemo-mechanically damaged African hair, which showed mostly endocuticle remnants. This is most likely due to the different effects that chemical straightening has on the hair vs. multiple cycles of perming the hair.

A more focused look into roughness and friction on the cuticle surface can be found by comparing Caucasian, Asian, and African virgin and chemo-mechanically damaged hair (Fig. 5.7) and virgin and virgin treated hair (Fig. 5.8). Virgin hair was used as the baseline to compare variations in roughness and friction force against modified hair (chemo-mechanically damaged or virgin treated). A scan size of $10 \times 10\,\mu m^2$ is displayed. Above each AFM and FFM image are cross-sectional plots of the surface (taken at the accompanying arrows) corresponding to surface roughness or friction force, respectively. From the surface roughness images, the step heights of one or more cuticle edges can be clearly seen. Step heights range from approximately 0.3 to $0.5\,\mu m$.

If the surface is assumed to have Gaussian height distribution and an exponential autocorrelation function, then the surface can be statistically characterized by just two parameters: a vertical descriptor, height standard deviation σ, and a spatial descriptor, correlation distance β^* (Bhushan, 1999a, 2002). The standard deviation σ is the square root of the arithmetic mean of the square of the vertical deviation from the mean line. The correlation length can be referred to as the length at which two data points on a surface profile can be regarded as being independent, thus serving as a randomness measure (Bhushan, 1999a, 2002). Table 5.1 displays these roughness parameters for each ethnicity as a function of hair type (virgin, chemo-mechanically damaged, and virgin treated) (LaTorre and Bhushan, 2005a). Virgin hair was shown to generally have the lowest roughness values, with virgin treated hair closely resembling virgin hair. Chemo-mechanically damaged hair showed a significantly higher standard deviation of surface height. This variation is expected because of the non-uniformity of the mechanical and chemical damage that occurs throughout a whole head of hair as well as each individual fiber. This is in agreement with the images of chemo-mechanically damaged hair shown previously, where regions of intact cuticle and severe degradation of the surface are intermingled. The

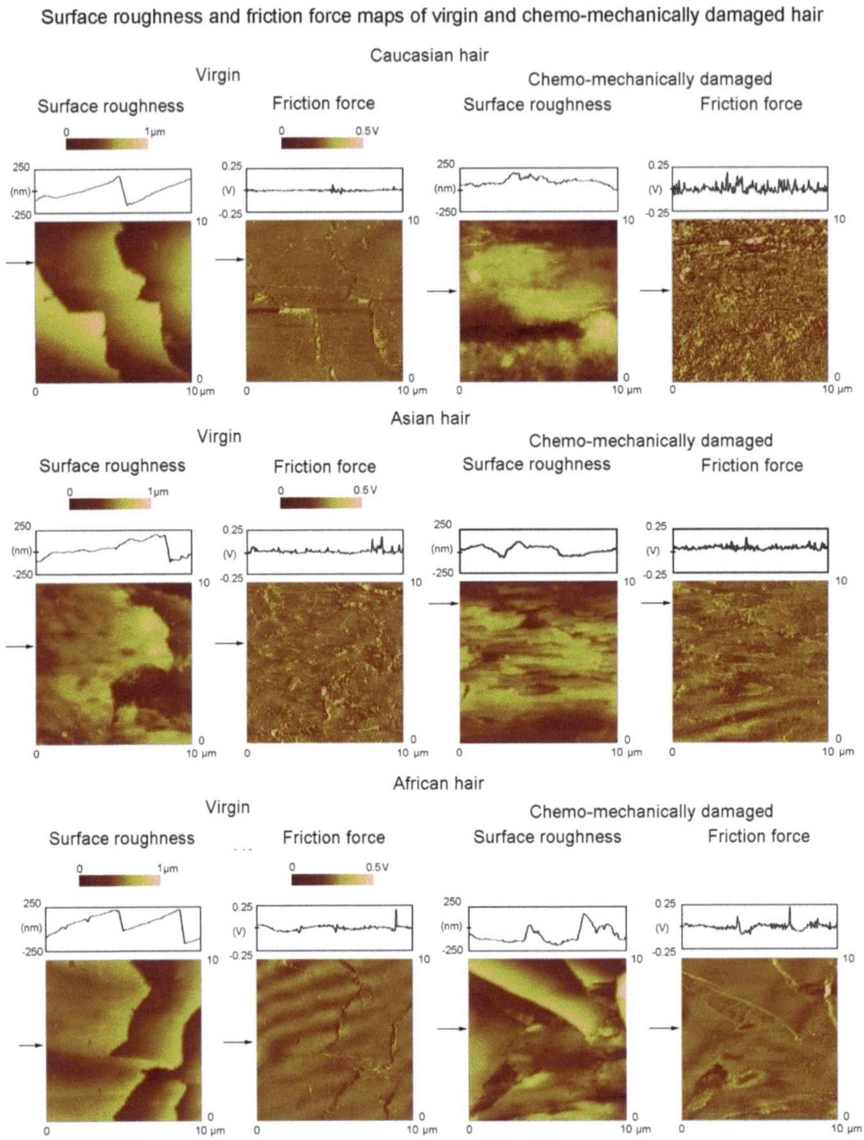

Fig. 5.7 Surface roughness and friction images for virgin and chemo-mechanically damaged Caucasian, Asian, and African hair at $10\,\mu\text{m}^2$ scan sizes. Shown above each image is a cross section taken at the corresponding arrows to show roughness and friction force information (LaTorre and Bhushan, 2005a)

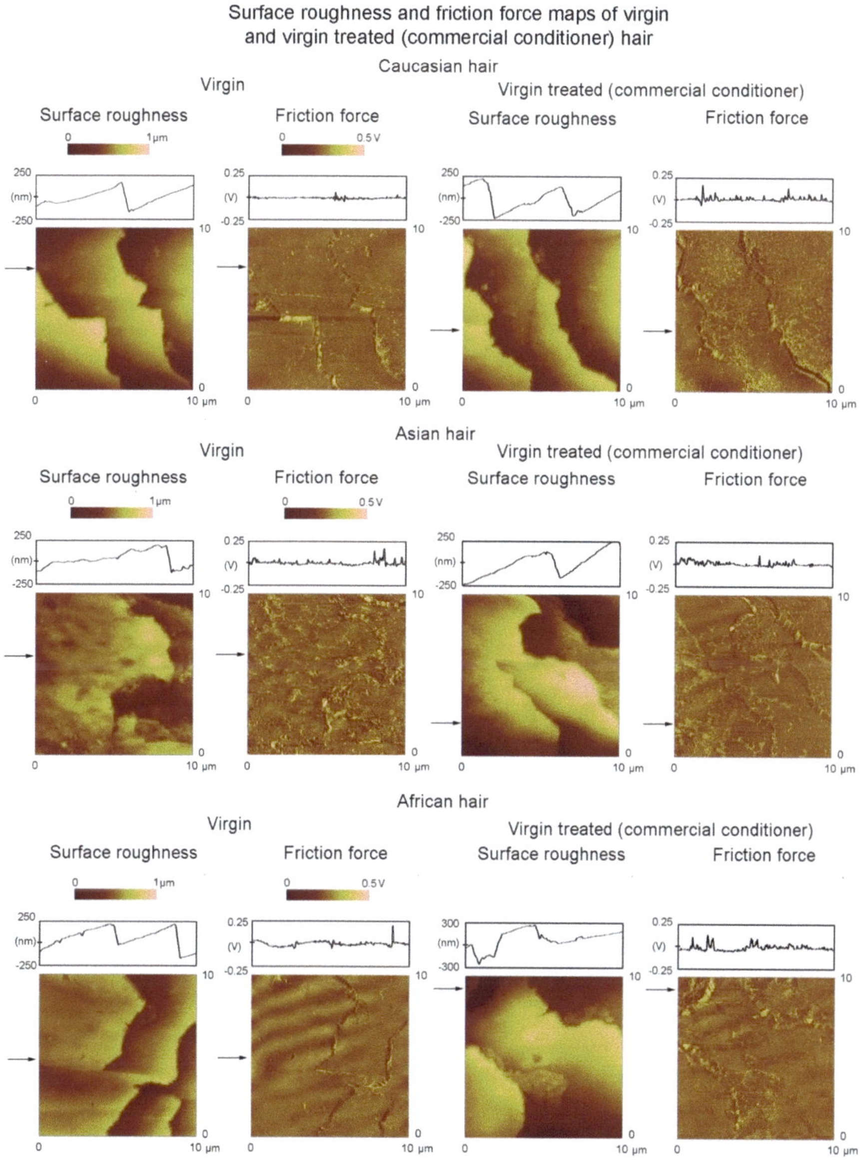

Fig. 5.8 Surface roughness and friction images for virgin and virgin treated Caucasian, Asian, and African hair at $10\,\mu m^2$ scan sizes. Shown above each image is a cross section taken at the corresponding arrows to show roughness and friction force information (LaTorre and Bhushan, 2005a)

Table 5.1 Surface roughness, coefficient of friction, and adhesive force values of virgin, chemo-mechanically damaged, and virgin treated (1 cycle commercial conditioner) hair at each ethnicity

	Virgin hair		Chemo-mechanically damaged hair		Virgin treated hair (commercial conditioner)	
	Surface roughness parameters					
	σ (nm)	β^* (μm)	σ (nm)	β^* (μm)	σ (nm)	β^* (μm)
Caucasian	12 ± 8	0.61 ± 0.3	17 ± 10	1.0 ± 0.3	12 ± 4	0.90 ± 0.3
Asian	9.7 ± 4	0.73 ± 0.3	33 ± 15	0.94 ± 0.3	7.1 ± 0.1	0.97 ± 0.3
African	12 ± 5	0.92 ± 0.3	21 ± 16	0.78 ± 0.3	11 ± 4	0.89 ± 0.2
	Coefficient of friction					
Caucasian	0.02 ± 0.01		0.13 ± 0.05		0.03 ± 0.01	
Asian	0.03 ± 0.01		0.13 ± 0.04		0.06 ± 0.04	
African	0.04 ± 0.02		0.14 ± 0.08		0.05 ± 0.01	
	Adhesive force (nN)					
Caucasian	25		16		32	
Asian	31		18		79	
African	35		38		63	

trends observed for standard deviation were not as evident for the correlation length β^*. For each ethnicity, chemo-mechanically damaged and virgin treated hair showed similar β^* values.

From Fig. 5.7, friction forces are generally seen to be higher on chemo-mechanically damaged hair than on virgin hair. Although friction forces were similar in magnitude, it was observed that the friction force on the cuticle surface of chemo-mechanically damaged hair showed a much larger variance, which contributed to the higher friction values. Another contribution to the higher friction could be that the tiny peaks that developed after damage also create a ratchet mechanism on a nanoscale, which affects the friction between the AFM tip and the surface. These peaks could then add to the friction signal. The damage of the hair by chemical and mechanical means has shown high reproducibility in the lab in terms of structure alteration, which explains the similar friction properties, no matter the ethnicity for chemo-mechanically damaged hair. With virgin and virgin treated hair, however, it is unknown what prior mechanical damage and sun exposure the fibers have seen and they vary largely on the individuals. Thus, across ethnicity there is variability in friction force for that hair.

Perhaps the most notable difference between virgin and virgin treated hair fibers can be seen in the friction force mappings of Fig. 5.8. Although quite comparable in surface roughness, close examination of the virgin treated hair surface shows an increase in friction force, usually only surrounding the bottom edge of the cuticle. This was unlike virgin hair, where friction generally remained constant along the surface, and unlike chemo-mechanically damaged hair, where there was large variability which was random over the entire surface.

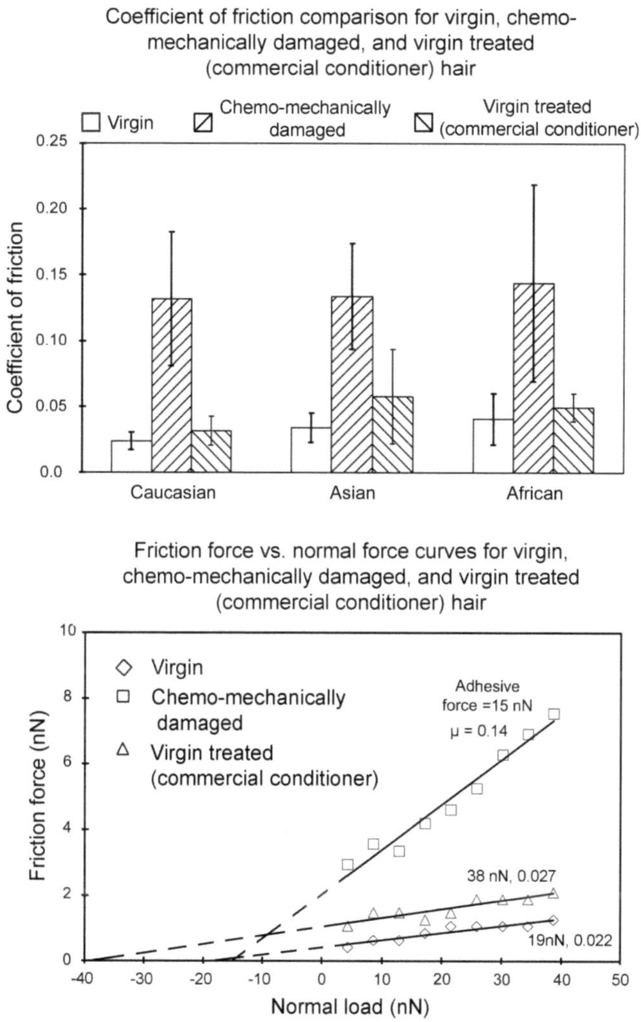

Fig. 5.9 Average coefficient of friction values for virgin, chemo-mechanically damaged, and virgin treated hair of each ethnicity. Error bars represent $\pm 1\sigma$ on the average coefficient value. In the *bottom* plot, friction force vs. normal load curves showing typical values for virgin, chemo-mechanically damaged, and virgin treated Caucasian hair. While the coefficient of friction is similar in virgin and treated hair, the adhesive force of treated hair is higher than in virgin hair (LaTorre and Bhushan, 2005a)

Figure 5.9 presents friction force curves as a function of normal load for Caucasian virgin, chemo-mechanically damaged, and virgin treated hair to further illustrate the previous discussion. One can see a relatively linear relationship between the data points for each type of hair sample. When plotted in such a way, the coefficient of friction is determined by the slope of the least squares fit line through the data. If this line is extended to intercept the horizontal axis, then a value for adhesive force can also be calculated, since friction force F is governed by the relationship

$$F = \mu(W + F_a) \tag{5.1}$$

where μ is the coefficient of friction, W is the applied normal load, and F_a is the adhesive force (Bhushan, 1999a, 2002).

One explanation for the increase in friction force of virgin treated hair on the micro/nanoscale is that during tip contact meniscus forces between the tip and the conditioner/cuticle become large as the tip rasters over the surface, causing an increase in the adhesive force. This adhesive force is of the same magnitude as the normal load, which makes the adhesive force contribution to friction rather significant. Thus, at sites where the conditioner is accumulated on the surface (namely around the cuticle scale edges), friction force actually increases. On the macroscale, however, the adhesive force is much lower in magnitude than the applied normal load, so the adhesive force contribution to friction is negligible over the hair swatch. As a result, virgin treated hair shows a decrease in friction force on the macroscale, which is opposite the micro/nanoscale trend. The friction and adhesion data on the micro/nanoscale are useful though, because they relate to the presence of conditioner on the cuticle surface and allow for obtaining an estimate of conditioner distribution.

It is also observed from Fig. 5.9 that while chemo-mechanically damaged hair displays a higher friction force on the application of normal load, and consequently a higher coefficient of friction, chemo-mechanically damaged hair friction is not as strongly dependent on the adhesive force contribution as the virgin and virgin treated hair. Average values for μ were calculated and are compiled in Figs. 5.9 and 5.10 and Table 5.1 for all hair ethnicities and types (LaTorre and Bhushan, 2005a). Error bars represent $\pm 1\sigma$ on the average coefficient value. The coefficients of friction for virgin, chemo-mechanically damaged, and virgin treated Caucasian hair are 0.02, 0.13, and 0.03, respectively. For virgin, chemo-mechanically damaged, and virgin treated Asian hair, coefficients of friction are 0.03, 0.13, and 0.06, respectively. Finally, virgin, chemo-mechanically damaged, and virgin treated African hair coefficients of friction are 0.04, 0.14, and 0.05, respectively. Chemo-mechanically damaged hair presents the highest coefficient of friction, but also displays the largest standard deviation, due to the large variations in chemical and mechanical damage that each hair or hair bundle experiences. Coefficient of friction of virgin treated hair is slightly larger than that of virgin hair for all ethnicities. While the coefficient of friction is similar in virgin and virgin treated hair, the adhesive force contribution to friction for Caucasian virgin treated hair is higher than in Caucasian virgin hair, when calculated according to the method described above. However, this was not always the trend for Asian and African virgin treated hair samples. It should be noted that since in friction force measurement the tip moves laterally over the surface, this might cause a smearing out of the conditioner layer which accounts for the inconsistent trend. In the adhesive force mappings described in the next section, where

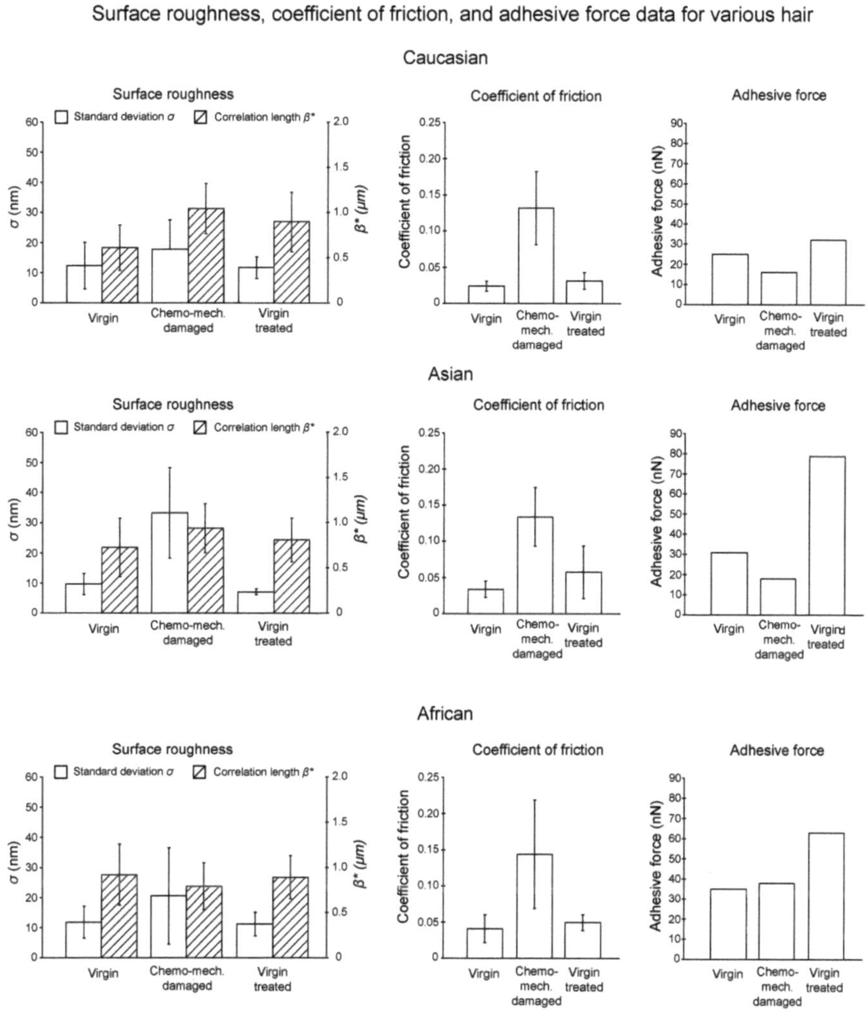

Fig. 5.10 Surface roughness, coefficient of friction, and adhesive force data for virgin, chemo-mechanically damaged, and virgin treated hair at each ethnicity (LaTorre and Bhushan, 2005a)

determination of adhesive force does not depend on this lateral movement, all virgin treated hair samples showed higher adhesion than their virgin hair counterparts.

A force calibration plot (FCP) technique and resulting adhesive force maps, commonly called force–volume (FV) maps, can be used to understand the adhesive forces between the AFM tip and the sample (Bhushan, 1999b, 2008b; Bhushan and Dandavate, 2000; Liu and Bhushan, 2003). Shown in Fig. 5.11 are FV maps and an example of the individual force calibration plots from which the maps were created. Adhesive force distribution for chemo-mechanically damaged hair was shown to be comparable to virgin hair adhesive force values, but slightly lower. A significant increase in adhesive force over the entire mapping was found in all cases

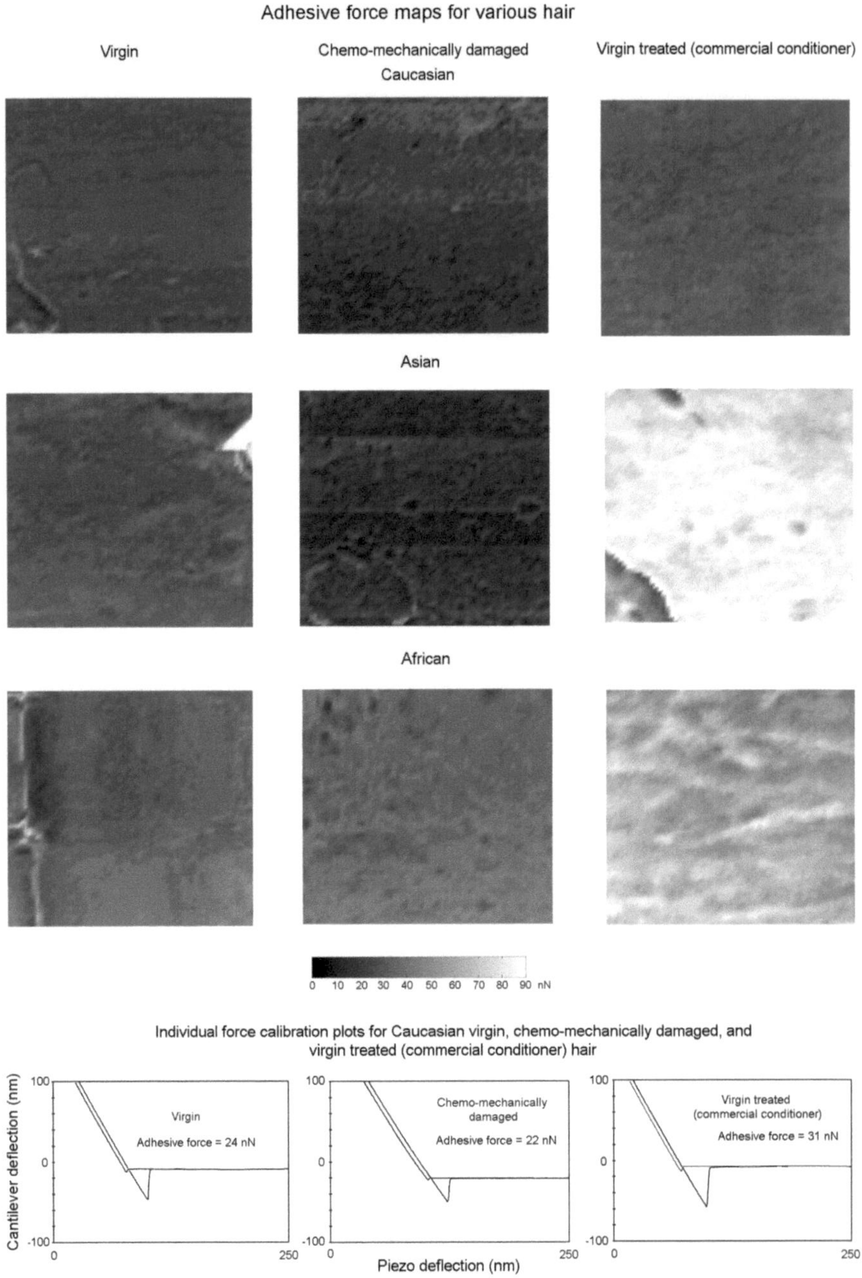

Fig. 5.11 Force–volume maps of virgin, chemo-mechanically damaged, and virgin treated hair at each ethnicity. Examples of the individual force calibration plots, which make up the FV maps, are presented for Caucasian hair of each type (LaTorre and Bhushan, 2005a)

of virgin treated hair as compared to virgin hair, especially in Asian and African hair. Conditioner distribution can be seen from these images. This technique shows promise to be very useful in further study of the distribution of materials and hair care products on the surface of the hair.

A typical value for the adhesive force of each FV map was calculated. Values are shown in the plot of Fig. 5.11, along with surface roughness and coefficient of friction data for all hair samples. Adhesive force values are also tabulated in Table 5.1.

5.2.1.1 Directionality Effects of Friction on the Micro/nanoscale

The outer surface of human hair is composed of numerous cuticle scales running along the fiber axis, generally stacked on top of each other. As previously discussed, the heights of these step changes are approximately 300 nm. These large changes in topography make the cuticle an ideal surface for investigating the directionality effects of friction using AFM/FFM.

The directionality effect of friction on the macroscale has been well studied in the past. It was shown by Robbins (1994) and Bhushan et al. (2005) that rubbing the hair from the tip to the root (against the cuticle steps) results in a higher coefficient of friction than rubbing the hair from root to tip (with the cuticles). On the micro/nanoscale, it is important to distinguish how material effects and topography-induced effects contribute to the directionality effect of friction force when scanning over and back across a small surface region (Bhushan, 1999a, b, 2002, 2008b). Figure 5.12 shows surface roughness, friction force, and surface slope maps of a Caucasian virgin hair fiber, each coupled to their accompanying 2D cross-sectional profiles. The scan size of $5 \times 5\ \mu m^2$ provides one cuticle step height to be studied. As the tip rasters over the step in the trace mode (i.e., from left to right), a small decrease in the friction force is observed as the tip follows the step downward. When the tip comes back in the retrace mode (i.e., right to left), climbing up the sharp peak results in a high friction signal. However, because of the sign convention of the AFM/FFM that causes a reversal in the sign when traversing the opposite direction, this signal is now observed to be highly negative. The interesting difference between the two profiles lies in the fact that the magnitude of the decrease in friction when going up the step is much larger than the magnitude of the friction when the tip is going down the step, yet both signals are in the same direction. The important result is that even by subtracting the two signals (T–R), there is still a gross variation in the image due to topography effects. These topography effects yield friction variations in the same direction, whereas material effects show up in opposite directions. It can be shown that the cuticle edge provides a local ratchet and collision mechanism that increases the friction signal at that point. It was concluded that surface slope variation always correlates to friction force variations with respect to topography effects, and the data presented in Fig. 5.12 show the same trend when comparing trace and negative retrace slope profiles.

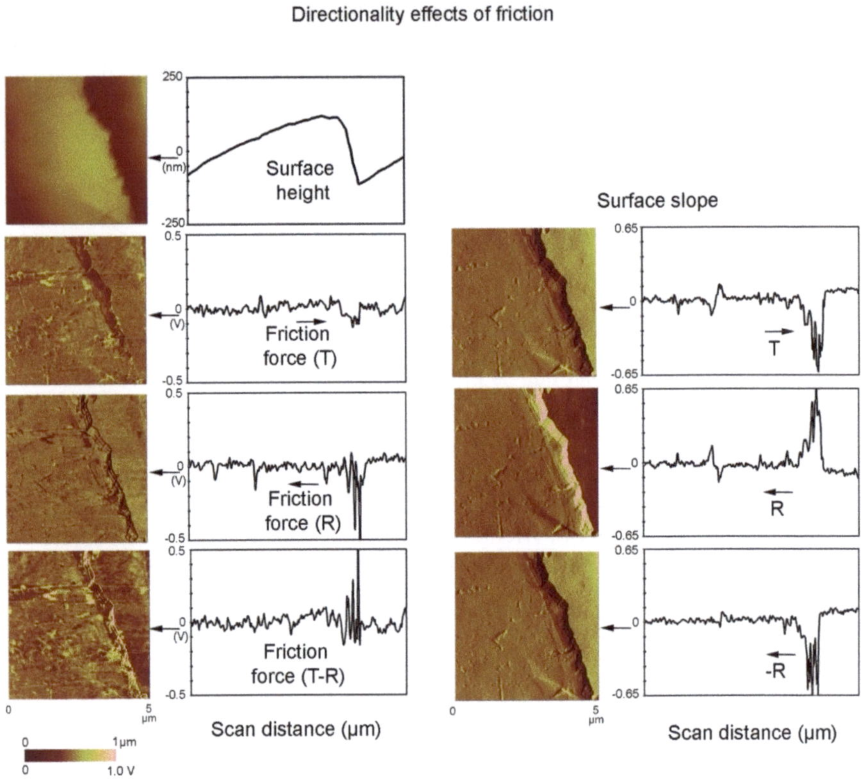

Fig. 5.12 Surface roughness, friction force, and slope across a cuticle scale edge of a Caucasian virgin hair to show the directionality dependence of friction. *Left*: Surface roughness and friction force mappings with accompanying 2D profiles. *Right*: Surface slope mappings with accompanying 2D profiles (LaTorre and Bhushan, 2005a)

5.2.2 Virgin and Chemically Damaged Caucasian Hair (With and Without Commercial Conditioner Treatment)

The hair surface is negatively charged and can be damaged by a variety of chemical (permanent hair waving, chemical relaxation, coloring, bleaching) and mechanical (combing, blowdrying) factors (Bolduc and Shapiro, 2001; Gray, 2001; Robbins, 1994). Figure 5.13 shows the transformation and wear of the cuticle scales before and after damage. Chemical damage causes parts of the scales to fracture and reveal underlying cuticle remnants. Conditioner application provides a protective coating to the hair surface for prevention of future damage.

Shown in Fig. 5.14 are surface roughness and friction force plots for virgin, virgin treated, chemically damaged, chemically damaged treated (one cycle conditioner), and chemically damaged treated (three cycles conditioner). Above each AFM and FFM image are cross-sectional plots of the surface (taken at the accompanying arrows) corresponding to surface roughness and friction force,

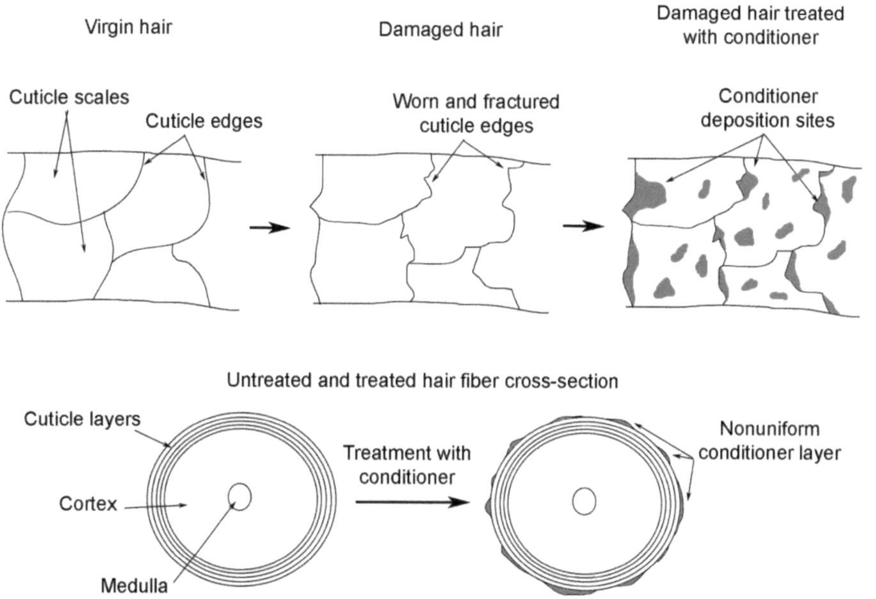

Fig. 5.13 Schematic of the effect of damage to the cuticle scales and the deposition of conditioner on the cuticle surface. The cross section of the hair with and without conditioner is shown below (LaTorre and Bhushan, 2006)

respectively. Although virgin and virgin treated hair are quite comparable in surface roughness maps, examination of the treated hair surface shows an increase in friction force, especially in the area surrounding the scale edge bottom level. These frictional patterns observed in treated hair were not like anything observed in the virgin or chemically damaged cases. Images of all hair types have shown friction variation due to edge contributions and cuticle mechanical damage that has left only remnants of cuticle sublayer (such as the endocuticle). Further investigation of the corresponding treated hair roughness images showed this increase in friction was not due to a significant change in surface roughness, either. One explanation for the increase in friction force of treated hair on the micro/nanoscale is that during tip contact, meniscus forces between the tip and the conditioner/cuticle become large as the tip rasters over the surface, causing an increase in the adhesive force. This adhesive force is of the same magnitude as the normal load, which makes the adhesive force contribution to friction rather significant. Thus, at sites where conditioner is accumulated on the surface (namely around the cuticle scale edges), friction force actually increases. On the macroscale, however, the adhesive force is much lower in magnitude than the applied normal load, so the adhesive force contribution to friction is negligible over the hair swatch. As a result, treated hair shows a decrease in friction force on the macroscale, which is opposite of the trend on the micro/nanoscale.

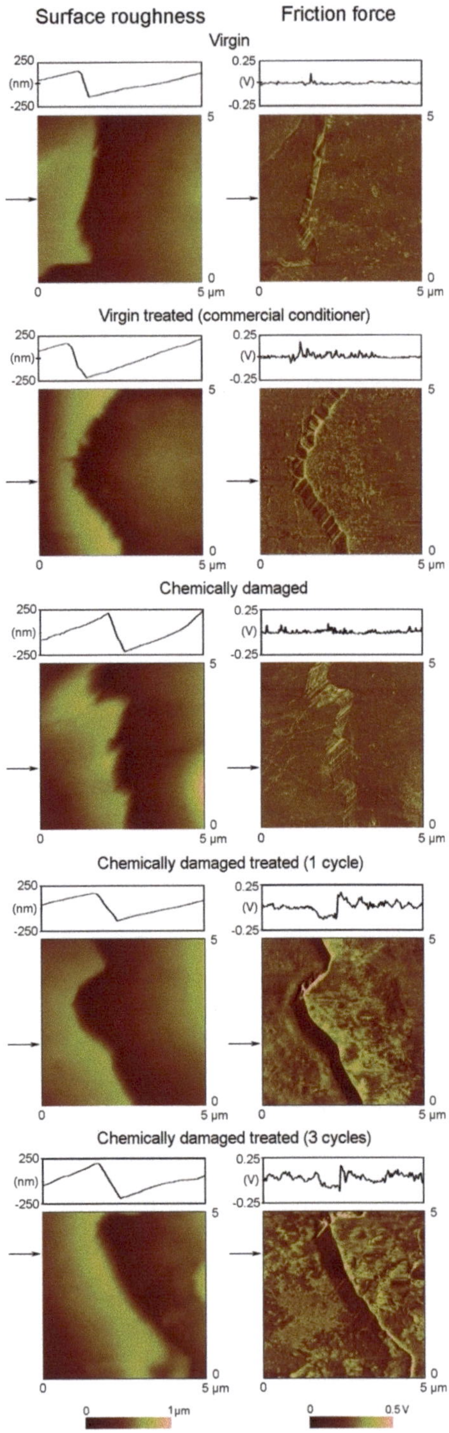

Fig. 5.14 Surface roughness and friction force images for Caucasian virgin, virgin treated, chemically damaged, damaged treated (1 cycle conditioner), and damaged treated (3 cycles conditioner) hair at 5 μm scan sizes. Shown above each image is a cross section taken at the corresponding arrows to show roughness and friction force information (LaTorre and Bhushan, 2005b)

In general, friction forces are higher on chemically damaged hair than on virgin hair. Although friction forces were similar in magnitude, it was observed that the friction force on the cuticle surface of chemically damaged hair showed a much larger variance, which contributed to the higher friction values. Chemically damaged treated hair shows a much stronger affinity to the conditioner. It is widely known that the cuticle surface of hair is negatively charged. This charge becomes even more negative with the application of chemical damage to the hair. As a result, the positively charged particles of conditioner have an even stronger attraction to the chemically damaged surface, which explains the increased presence of conditioner (and corresponding higher friction forces) when compared to virgin treated hair. With the application of three conditioner cycles on chemically damaged treated hair, friction force is still higher near the cuticle edge; however, it is also increased all over the cuticle surface, showing a more uniform placement of the conditioner.

Figure 5.15 shows adhesive force maps for the various hair, which gives a measurement of adhesive force variation on the surface. Figure 5.16 presents surface roughness, coefficient of friction, and adhesive force plots for the various virgin and chemically damaged hair discussed above. The data are also presented in Table 5.2 (LaTorre and Bhushan, 2005b). Surface roughness for human hair is characterized by a vertical descriptor, height standard deviation σ, and a spatial descriptor, correlation distance β^* (Bhushan, 1999a, 2002). The standard deviation σ is the square root of the arithmetic mean of the square of the vertical deviation from the mean line. The correlation length can be referred to as the length at which two data points on a surface profile can be regarded as being independent, thus serving as a randomness measure. These two parameters are all that is needed if the surface is assumed to have Gaussian height distribution and an exponential autocorrelation function (Bhushan, 1999a, 2002). Virgin and virgin treated hair showed similar σ values, while β^* was higher in virgin treated hair. Chemically damaged hair and both types of chemically damaged treated hair showed similar roughness values, although σ was higher for the treated cases. The chemically damaged hair presented in this work is different from the chemo-mechanically damaged hair studied by LaTorre and Bhushan (2005a). It seems that chemically damaging the surface does not lead to as much wear and surface roughness increase as the combination of both chemical and mechanical damage. Thus, it should be noted (and it is understandable) that there are differences between the chemo-mechanically and chemically damaged hair.

Coefficient of friction of virgin and virgin treated hair is similar, but slightly higher for the treated cases. Chemically damaged hair shows a much higher coefficient of friction, with more variation in the values since the chemical damage varies throughout each individual fiber. An interesting finding was that, contrary to the virgin and virgin treated hair results, coefficient of friction for chemically damaged hair decreased with the application of conditioner treatment (both one and three cycles). One possible explanation is that because the stronger negative charge on chemically damaged hair results in better attraction of conditioner, this leads to higher adhesive force but, more importantly, lower shear strength on the surface. This creates an overall effect of lubrication and ultimately lowers the coefficient of friction.

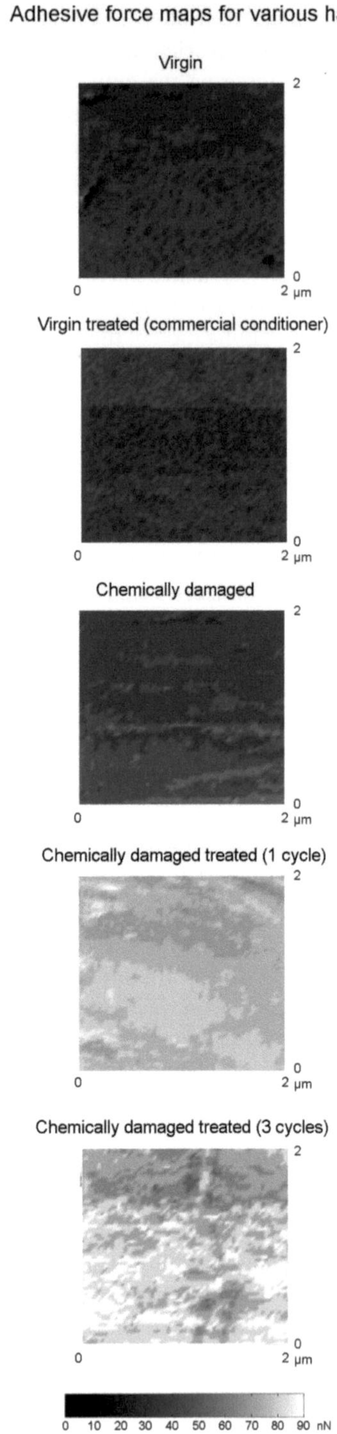

Fig. 5.15 Adhesive force maps displaying variations in adhesive force on the cuticle surface of Caucasian virgin, virgin treated, chemically damaged, damaged treated (1 cycle conditioner), and damaged treated (3 cycles conditioner) hair. Treated hair is shown to have higher adhesive force due to meniscus effects (LaTorre and Bhushan, 2005b)

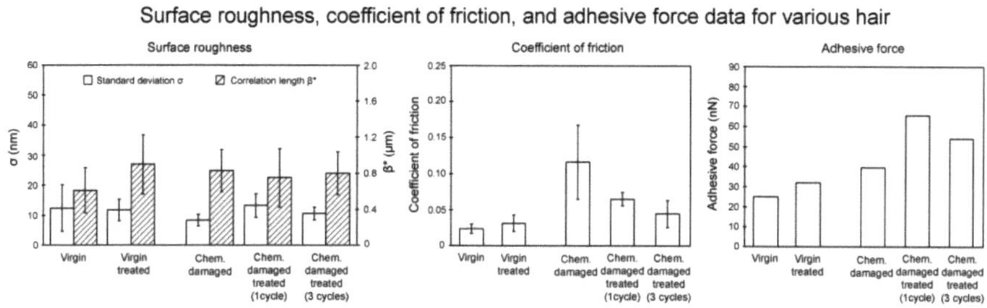

Fig. 5.16 Surface roughness, coefficient of friction, and adhesive force plots for Caucasian virgin, virgin treated, chemically damaged, damaged treated (1 cycle conditioner), and damaged treated (3 cycles conditioner) hair (LaTorre and Bhushan, 2005b)

Table 5.2 Surface roughness, coefficient of friction, and adhesive force for virgin and chemically damaged Caucasian hair, with and without conditioner treatment

	Surface roughness			
Hair type	σ (nm)	β^* (μm)	Coefficient of friction	Adhesive force (nN)
Virgin	12 ± 8	0.61 ± 0.2	0.02 ± 0.01	25 ± 5
Virgin, treated	12 ± 4	0.90 ± 0.3	0.03 ± 0.01	32 ± 5
Chemically damaged	8.4 ± 2	0.83 ± 0.2	0.13 ± 0.06	39 ± 0.5
Damaged, treated (1 cycle)	13 ± 4	0.75 ± 0.3	0.05 ± 0.02	66 ± 0.7
Damaged, treated (3 cycles)	11 ± 2	0.80 ± 0.2	0.04 ± 0.02	54 ± 33

Average adhesive force values were taken from the adhesive force maps described previously. Virgin treated hair shows a higher adhesive force than virgin hair due to the meniscus effects that come about from AFM tip interaction with the conditioner on the cuticle surface (Bhushan, 1999a, 2002). The same trend is true and even more evident for chemically damaged treated hair compared to chemically damaged hair. A possible reason that one cycle of conditioner on chemically damaged hair showed higher average adhesive force than three cycles could be because the three cycles treatment generally places the conditioner more uniformly on the surface rather than accumulating it most near the bottom surface near the cuticle edge, which is where the adhesive force maps were generally taken. Nevertheless, the increased adhesive force shown in the plots is a clear indication of conditioner present on the hair surface, and its localization can be observed.

5.2.2.1 Effect of Relative Humidity, Temperature, Soaking, and Durability Measurements

Figure 5.17 displays the effect of relative humidity on friction force and adhesive force. Coefficient of friction remained relatively constant for virgin and virgin treated hair. However, chemically damaged hair experienced a large increase in coefficient of friction at high humidity, while chemically damaged treated hair experienced the opposite trend. This clearly shows that heavy moisture in the air plays a role on the frictional properties of chemically damaged hair. When combined

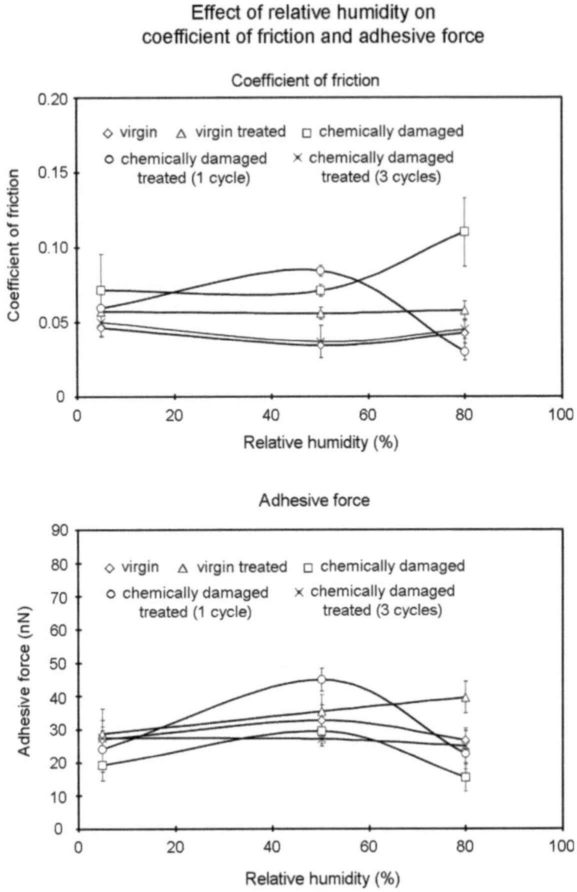

Fig. 5.17 Effect of relative humidity on nanotribological properties of Caucasian virgin, virgin treated, chemically damaged, damaged treated (1 cycle conditioner), and damaged treated (3 cycles conditioner) hair (LaTorre and Bhushan, 2005b)

with conditioner, a lubricating effect once again dominates as the water helps form a liquid layer which is more easily sheared. In terms of adhesive force, most samples showed a decrease in adhesive force with high humidity. It is expected that as water builds up on a surface, meniscus effects diminish and as a result do not readily contribute to adhesive force. Thus, the adhesive force is expected to come down at very high humidity.

Figure 5.18 displays the effect of temperature on friction force and adhesive force. The coefficient of friction generally decreased with increasing temperature. As the hair fiber heats up, conditioner which is present on the surface decreases in viscosity, causing a thinner film and lower friction force. The lower friction force ultimately leads to lower coefficient of friction values. Adhesive force was shown to decrease with increasing temperature as well. This was especially evident for treated

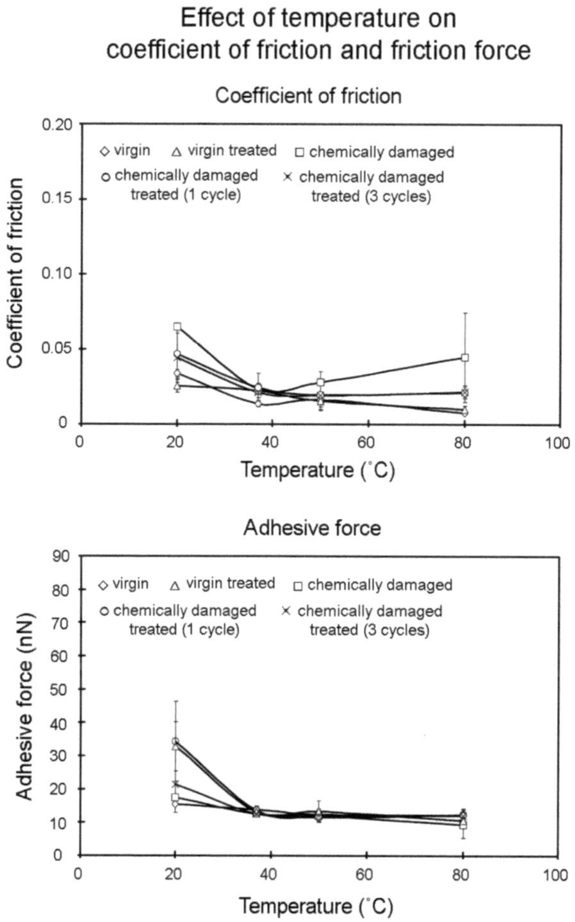

Fig. 5.18 Effect of temperature on nanotribological properties of Caucasian virgin, virgin treated, chemically damaged, damaged treated (1 cycle conditioner), and damaged treated (3 cycles conditioner) hair (LaTorre and Bhushan, 2005b)

hair fibers, whereas large adhesive force at room temperature decreased rapidly to adhesion values similar to non-treated fibers. It is most likely that at higher temperatures the thinning conditioner layer causes a reduced surface tension, which directly relates to the drop in adhesive force.

Virgin, chemically damaged, and chemically damaged treated hair samples were soaked in de-ionized water for 5 min. Their corresponding coefficient of friction was measured and compared to coefficient of friction values for dry samples which were adjacent to the wet samples on the respective hair fiber. Figure 5.19 shows the results for two hair samples of each hair type. Virgin hair exhibits a decrease in coefficient of friction after soaking. Virgin hair is more hydrophobic (see Table 2.4), so more of the water is present on the surface and results in a lubrication effect

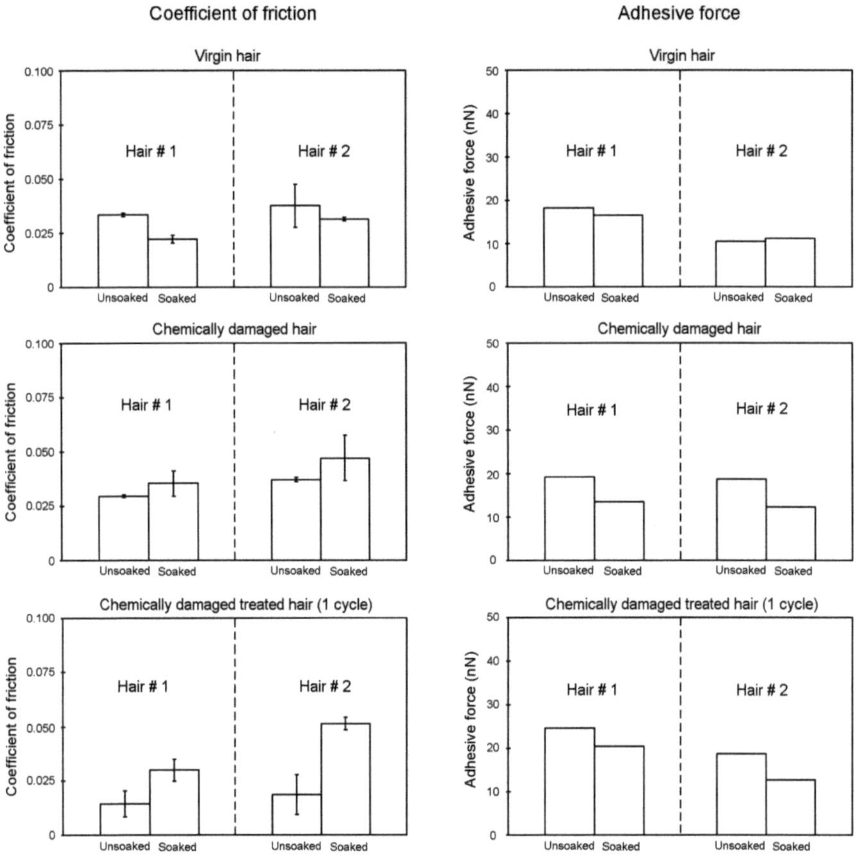

Fig. 5.19 Effect of soaking in de-ionized water on coefficient of friction and adhesive force for Caucasian virgin, chemically damaged, and chemically damaged treated (1 cycle conditioner) hair (LaTorre and Bhushan, 2006)

after soaking. Chemically damaged hair tends to be hydrophilic, due to the chemical degradation of the cuticle surface, and results in absorption of water after soaking. This softens the hair, which leads to higher friction, even with conditioner treatment. This is yet another indication that virgin and chemically damaged hair have significantly different surface properties which in many cases results in opposite trends for their nanoscale tribological properties. Adhesive force for virgin hair remained approximately the same before and after soaking, while it decreased for chemically damaged and chemically damaged treated hair after soaking.

Figure 5.20 shows the durability effects on friction force for various hair. Above the graph are pictures of unworn and worn virgin hair, with the cuticle edge serving as a reference point. Before testing, the surface is relatively smooth and void of any large debris or wear. After 1000 cycles at approximately 10 μN load with a stiff silicon AFM tip, however, the interaction has caused degradation and wear (scratch) marks on the cuticle scale. This is the type of wear one could potentially

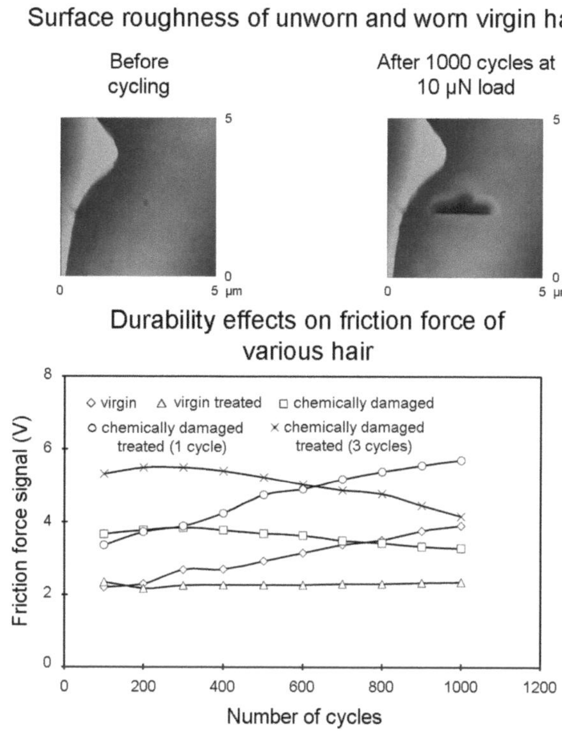

Fig. 5.20 Durability study of friction force change as a function of AFM tip cycling for Caucasian virgin, virgin treated, chemically damaged, damaged treated (1 cycle conditioner), and damaged treated (3 cycles conditioner) hair. The images *above* the plot signify before and after comparisons of a cuticle surface subjected to cycling at a 10 μN load (LaTorre and Bhushan, 2005b)

see if hair were to come in contact with sand from a day spent at the beach, among other activities. Virgin hair shows an obvious increase in friction force signal as the scratch mark digs further into the surface. By this time the lubricious lipid layer on the surface of the virgin cuticle has been worn away and the friction force comes close to the magnitude of chemically damaged hair friction force at the onset of cycling. When conditioner is applied to the virgin hair, however, the wear does not show up as an increase in the friction force. Thus, conditioner serves as a protective covering to the virgin hair and helps protect the tribological properties when wear ensues.

5.2.3 *Various Hair Types Treated with Various Conditioner Matrices*

LaTorre and Bhushan (2006) measured the nanotribological properties of various conditioner matrices and compared their performance with a commercial conditioner. As mentioned earlier, a conditioner primarily consists of cationic surfactants, fatty alcohols, and PDMS blend silicone (dimethicone) (Table 1.7). They studied two different silicones – a PDMS (blend of low and high MW) silicone and an amino silicone added to BTMAC surfactant. Motivation for selection of amino silicone was that it may attach to the damaged hair surface and improve nanotribological performance and durability.

Figure 5.21 displays the representative surface roughness and friction force maps for chemically damaged hair, commercially treated hair, PDMS blend silicone treated hair, and amino silicone treated hair (LaTorre and Bhushan, 2006). When conditioner is applied to the surface, a pattern of high friction is shown in the area surrounding the bottom edge of the cuticle. This is believed to be an area of conditioner accumulation which causes increased friction due to meniscus effects. Friction maps for PDMS blend silicone do not show this increase as readily, suggesting that this type of silicone is not a contributor of high friction force on the nanoscale. This can be due to the fact that a PDMS-type silicone is fairly mobile on the surface and thus does not cause the same meniscus effects as the AFM tip rasters through it. The amino group typically is less mobile and harder to move, which accounts for a different slip plane flow than PDMS silicone.

Figure 5.22a displays the adhesive force maps for chemically damaged hair and the different treatments (LaTorre and Bhushan, 2006). As shown in the legend, a lighter area corresponds to a higher tip pull-off force (adhesive force). Chemically damaged untreated hair has relatively low adhesive force and is more or

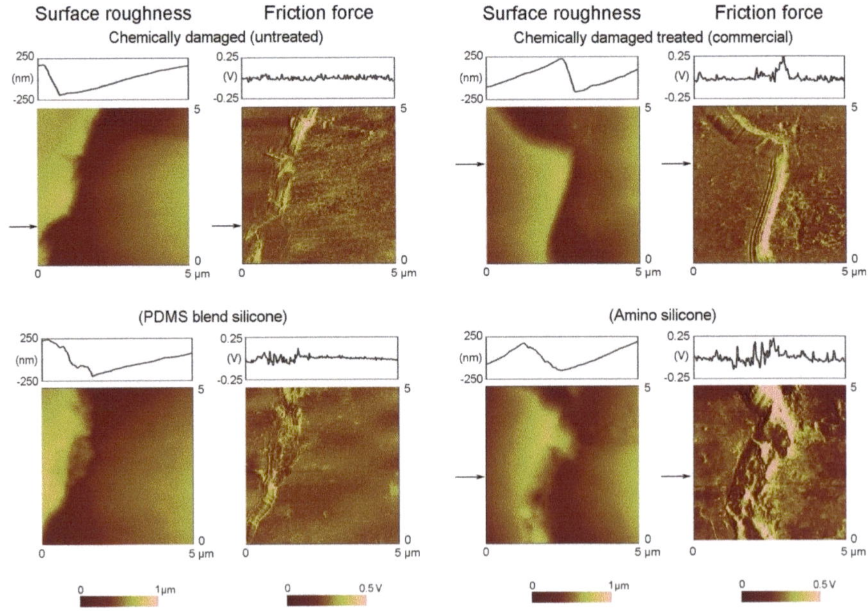

Fig. 5.21 Surface roughness and friction force maps of Caucasian chemically damaged hair with various treatments

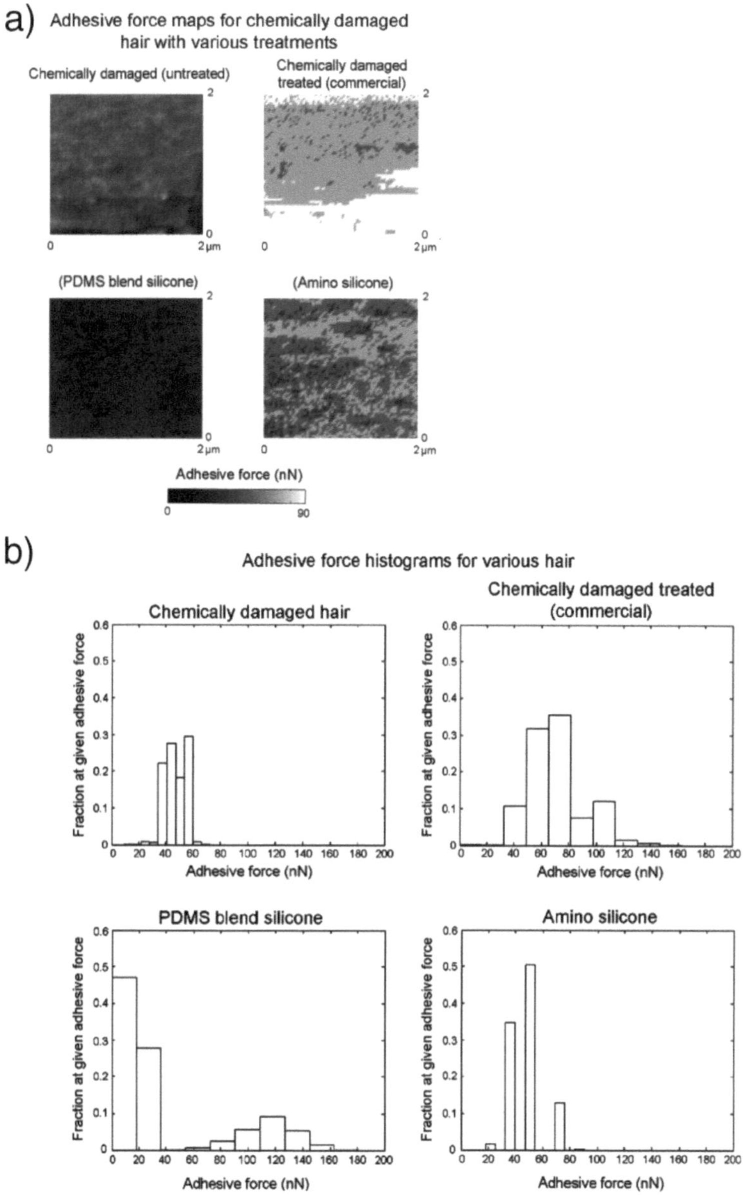

Fig. 5.22 (**a**) Adhesive force maps for Caucasian chemically damaged hair with various treatments; (**b**) adhesive force histograms (LaTorre and Bhushan, 2006)

less consistent over the hair surface. In nearly all cases, addition of a conditioner treatment caused an increase in meniscus forces, which in turn increased the adhesive pull-off force between the AFM tip and the sample. Observing the chemically damaged treated hair, the uneven distribution of the conditioner layer is seen. This uneven distribution is also most evident for the amino silicone images, in which the less mobile silicone brings about a distinguishable change in adhesion over the surface. For the PDMS blend silicone, it is seen that adhesion over the surface is much more consistent than the amino silicone, where various areas of high adhesion occur.

It is important to note that while adhesive force maps presented are representative images for each treatment, adhesive force varies significantly when treatments are applied to the hair surface. Figure 5.22b shows histograms of all adhesive force data for chemically damaged, chemically damaged treated, PDMS blend silicones, and amino silicones (LaTorre and Bhushan, 2006). Chemically damaged treated hair shows a much larger range of adhesive force values and a normal distribution, which suggests that the conditioner layer is normally distributed. The histogram for PDMS blend silicone treatment shows a normal distribution at the larger adhesive force values, but also shows another peak at low adhesion values. Amino silicone treated hair follows a normal distribution, but it is interesting to note the distinct groupings of the adhesion values and the spacing between them. This is further evidence that the amino silicone groups most likely attach immediately to the hair surface and are less mobile than PDMS silicone, causing distinct regions of high and low adhesion values over the cuticle surface.

Figure 5.23 displays a summary of the data collected for all chemically damaged hair samples and their treatments (LaTorre and Bhushan, 2006). The figure also includes the macroscale coefficient of friction data obtained using a technique similar to the macroscale measurement technique described earlier (Fig. 2.7). Table 5.3 reviews some of the observations and corresponding mechanisms which help to explain the trends found (LaTorre and Bhushan, 2006). The application of the commercial conditioner to the chemically damaged hair caused a decreased coefficient of friction and a large increase in adhesive force. The decreased coefficient of friction may be explained by the fact that the chemically damaged hair accumulates much of the positively charged conditioner on the surface due to its highly negative charge, which in turn makes it easier to shear the liquid on the surface, causing lower coefficient of friction. However, the nanoscale pull-off force (adhesive force) is much larger than on the untreated hair because of meniscus effects. In general, adhesive force varied widely, but typically showed a significant increase with the presence of conditioner. As discussed previously, this is a clear sign that meniscus effects are influencing the pull-off force between the tip and the sample.

In most cases, the macroscale and microscale coefficient of friction followed the same trend, in which a decrease was observed with the addition of the PDMS blend or amino silicones to the surfactant. The silicones are typically used as a major source of lubrication and thus give the conditioner more mobility on the hair surface compared to just surfactants and fatty alcohols. The inverse trend was seen only for the amino silicone group.

The amino silicones have a strong electrostatic attraction to the negatively charged hair surface, which in turn creates higher binding forces and less mobility. The dampened mobility of the amino silicone, with respect to hair surface and tip, may account for this wide variation in coefficient of friction and large adhesive

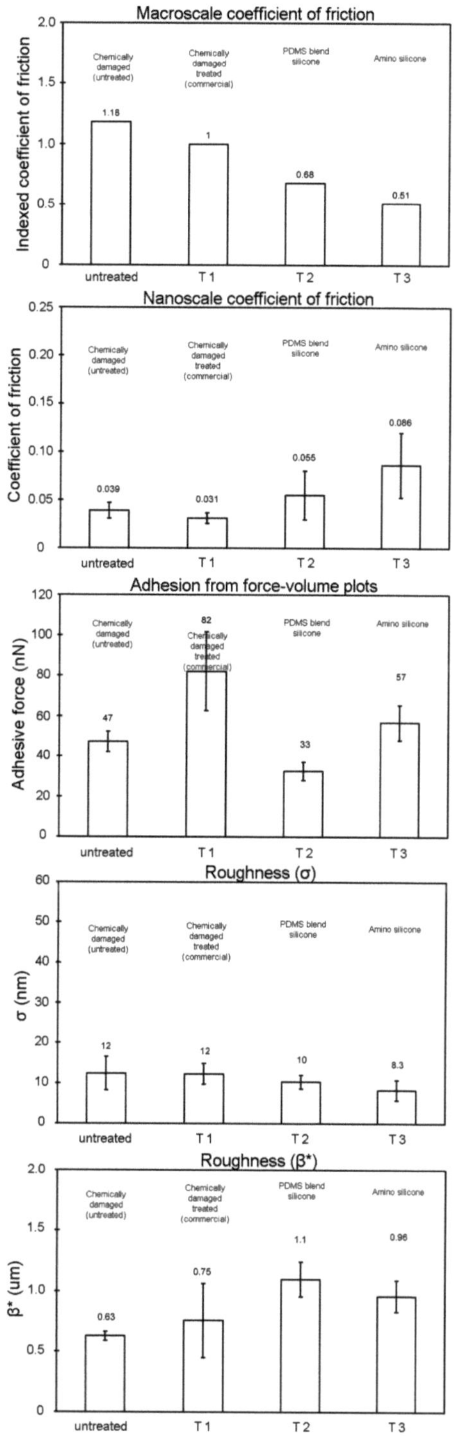

Fig. 5.23 Coefficient of friction, adhesive force, and surface roughness plots for Caucasian chemically damaged hair with various treatments labeled as T_1, T_2, and T_3

Table 5.3 Observations and corresponding mechanisms regarding coefficient of friction and adhesion for various hair treatments

Observation	Mechanism
Damaged vs. damaged treated hair	
Damaged hair shows a decrease in coefficient of friction but an increase in adhesion from the application of commercial conditioner	The conditioner layer deposited on the surface of the damaged hair results in a lower shear strength which in turn lowers the coefficient of friction, while meniscus effects increase the pull-off (adhesive) force between the tip and hair sample
PDMS blend vs. amino silicone	
Amino silicones interact strongly with negatively charged hair surface	A stronger electrostatic attraction exists which results in stronger binding forces (which leads to higher adhesion) for amino silicone
Amino silicone thickness distribution on hair is less uniform than with PDMS blend	Less mobility with amino silicones, so molecules attach to hair at contact and do not redistribute easily

force values. In terms of adhesive force, it was previously observed in Fig. 5.22a that the amino silicone treatments showed much more distinct regions of higher and lower adhesion compared to PDMS blend silicones. This non-uniform amino silicone thickness distribution on hair is also believed to be caused by the inhibited mobility, as the molecules immediately attach to the hair at contact and do not redistribute as a uniform coating. The increased polarity of the amino silicones compared to the PDMS blend can also be a major contributor of the higher friction and adhesion at high deposition levels.

In respect to roughness, the vertical standard deviation decreased slightly with most treatments, although standard deviations were similar. The spatial parameter increased slightly with treatments, but the variation also becomes extremely high.

5.2.4 Skin

Synthetic materials were also studied for surface roughness and friction force information, shown in Fig. 5.24a. While macroscale dimples could be seen on the surface of collagen film, it was interesting to find similar pits and dimples on the micro/nanoscale, consequently with a large variation in dimple size and depth. Polyurethane films are shown to have quite different topography and friction forces, while their coefficient of friction is very similar. Human skin shows a rougher texture with higher peaks; see Fig. 5.24b. The roughness parameters for the collagen and polyurethane films, and also for human skin, are presented in Table 5.4 (LaTorre and Bhushan, 2005a). Surface height standard deviation σ was approximately three times larger than that of virgin hair for both synthetic materials. However, the correlation length β^* was lower than what was typically observed in hair. The average

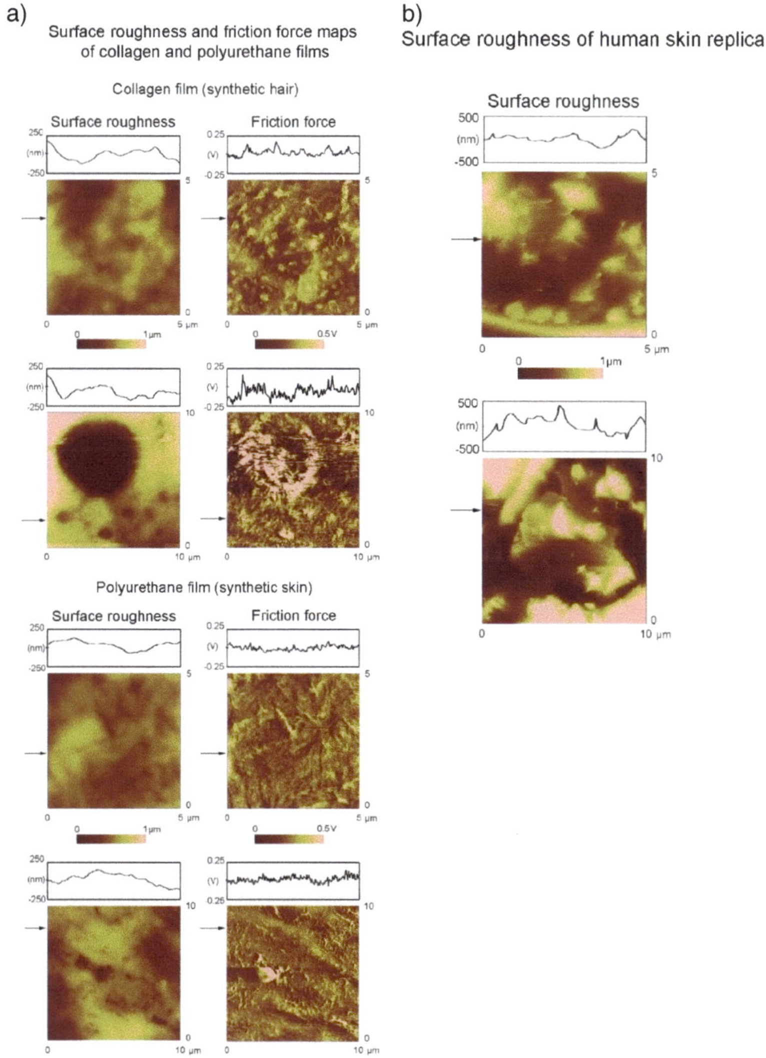

Fig. 5.24 (**a**) Surface roughness and friction force images for collagen and polyurethane films at 5 and 10 μm^2 scan sizes. Shown *above* each image is a cross section taken at the corresponding arrows to show roughness and friction force information; (**b**) surface roughness for human skin at 5 and 10 μm^2 scan sizes. Note that vertical scales in 2D section profiles in part (**b**) are doubled (LaTorre and Bhushan, 2005a)

Table 5.4 Surface roughness parameters σ and β^* for collagen and polyurethane films and human skin

	σ (nm)	β^* (μm)
Collagen (synthetic hair)	36 ± 11	0.50 ± 0.1
Polyurethane (synthetic skin)	33 ± 6	0.71 ± 0.1
Human skin	80 ± 28	0.59 ± 0.2

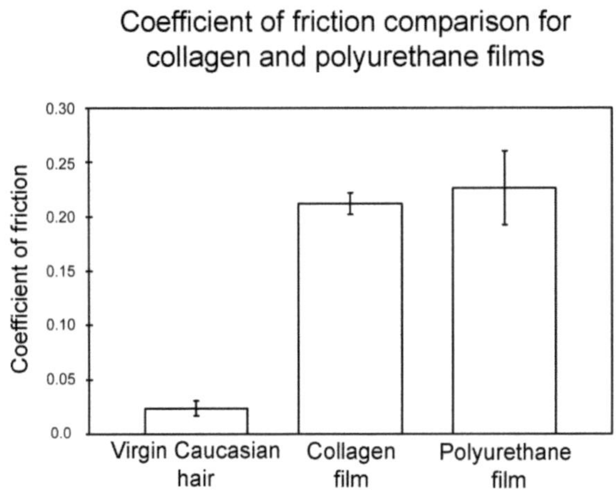

Fig. 5.25 Average coefficient of friction values for collagen and polyurethane films. Error bars represent $\pm 1\sigma$ on the average coefficient value (LaTorre and Bhushan, 2005a)

coefficient of friction for these synthetic materials is shown in Fig. 5.25 plotted next to virgin Caucasian hair as a reference. These values were calculated using the slope of the friction force curves, described previously. Both collagen and polyurethane films displayed similar coefficient of friction values of 0.22 and 0.24, respectively. Virgin hair displays a much lower coefficient of friction than both materials, approximately eight times lower.

5.3 Scale Effects

5.3.1 Directionality Dependence of Friction

Human hair has been shown in the past to have a directionality friction effect on the macroscale, making it easier to travel over the hair surface from root to tip than in the opposite direction due to the tile-like orientation of cuticles (Bhushan et al., 2005; Robbins, 1994). The outer surface of human hair is composed of numerous cuticle scales running along the fiber axis, generally stacked on top of each other. As previously discussed, the heights of these step changes are approximately 300 nm. These large changes in topography make the cuticle an ideal surface for investigating the directionality effects of friction (LaTorre and Bhushan, 2006). The first row of Fig. 5.26a displays a low-resolution SEM micrograph and a friction profile for macroscale coefficient of friction measurements. Note that these data are taken for measurements where multiple fibers are in contact with the synthetic skin upper specimen at the same time and do not correspond directly with the SEM micrograph. With an applied normal load of 50 mN and 3 mm travel, it is observed that the friction force produced when scanning from root to tip (referred to as "along cuticle") is lower than that when scanning against the cuticle. This is a direct consequence of the literally thousands of scale edges which come in contact with the synthetic skin. When traveling against the cuticle, these edges act as tiny resistors to motion as they are forced backward and uplifted from their interface with the underlying cuticle layers. The resistance to motion of so many cuticle edges at the same time becomes "additive" and results in higher values of friction, corresponding to a higher coefficient of friction than when traveling along the cuticle. For along-cuticle travel, these edges are forced downward against the underlying cuticle layers so that the resistance effect of these edges is limited, which results in lower friction values.

The second row of Fig. 5.26a shows an AFM height map corresponding to the microscale friction profile shown to its right. (These measurements were made using an AFM tip mounted with 4 μm radius silica ball.) Due to the size of the hair, it was only possible to capture a rectangular height map as shown. It is evident that the 100 μm travel results in the involvement of several cuticle scales. The applied normal load for the microscale friction profile (about 20 nN) is significantly reduced from that of the macroscale value, which consequently yields much lower friction forces. In the along-cuticle direction, we can see small fluctuations in the friction data over the scan distance. These are caused by local variations in surface roughness, system noise, and changes due to traveling over the scale edges. However, when scanning against the cuticle, distinctly large spikes in the data are observed at roughly 5–10 μm intervals. This is clearly the effects of the scale edges coming in contact with the AFM microscale tip and causing local collisions and ratcheting of the tip. Because of the sign convention of the AFM that causes a reversal in the sign when traversing the opposite direction, this signal is now observed to be highly negative. These edge effects are hence the primary item responsible for the higher friction and coefficient of friction observed in the against-cuticle direction.

Directionality effects on nanoscale hair friction have previously been reported by LaTorre and Bhushan (2005a). As shown in the bottom row of Fig. 5.26a, the scan size of $5 \times 2.5\,\mu m^2$ provides one cuticle scale edge to be studied (the height map is rectangular only to be consistent with the microscale image in the middle row). As the tip rasters in the along-cuticle direction, a small decrease in the friction force is observed as the tip follows the scale edge on a downward slope. When the tip comes back in the against-cuticle direction, colliding with the scale edge and climbing up the sharp peak result in a high friction signal. As discussed above, because of the sign convention of the AFM/FFM that causes a reversal in the sign when traversing the opposite direction, this signal is now observed to be highly negative. The interesting difference between the two profiles lies in the fact that the magnitude of the decrease in friction when going up the step is much larger than the

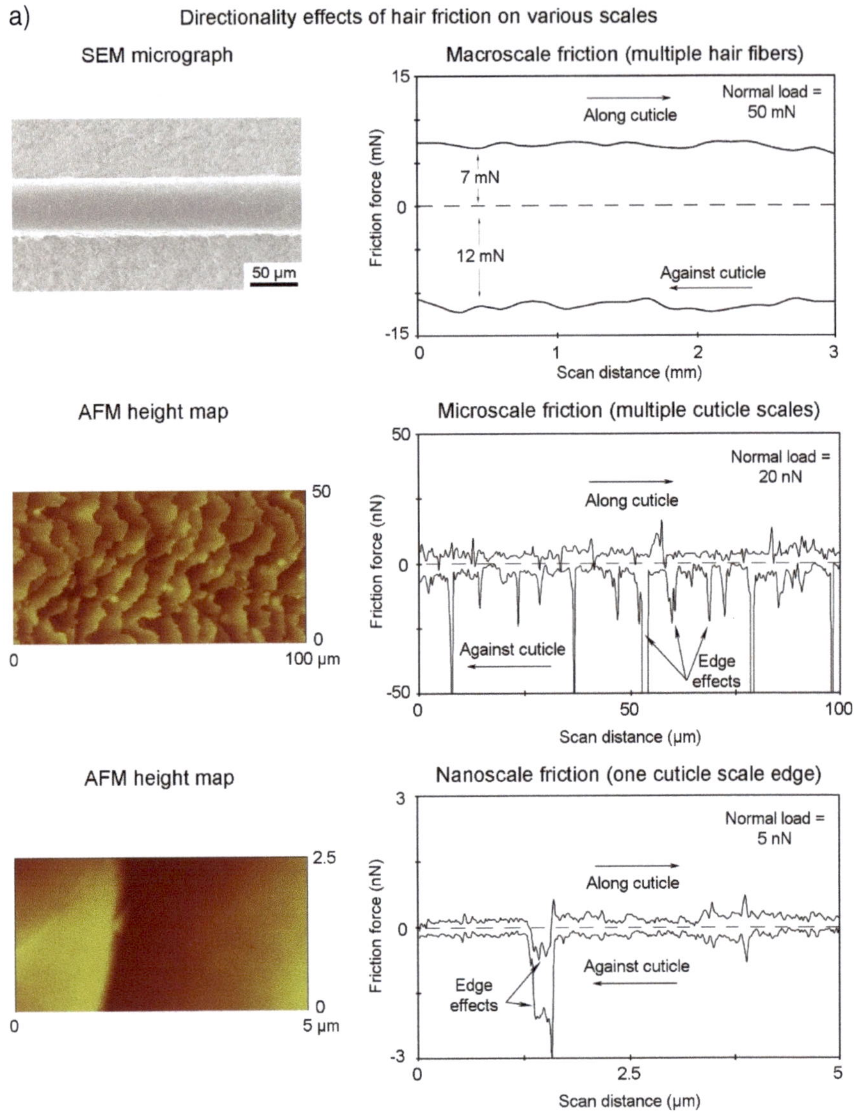

Fig. 5.26 (**a**) Directionality effects of virgin Caucasian hair friction on various scales; (**b**) microscale coefficient of friction data for Caucasian virgin, chemically damaged, virgin treated, damaged treated (1 cycle conditioner), and damaged treated (3 cycles conditioner) hair showing directionality effects (LaTorre and Bhushan, 2006)

magnitude of the friction when the tip is going down the step, yet both signals are in the same direction. It is thus shown that the cuticle edge provides a local ratchet and collision mechanism that increases the friction signal at that point and clearly shows the directionality dependence caused by edge effects.

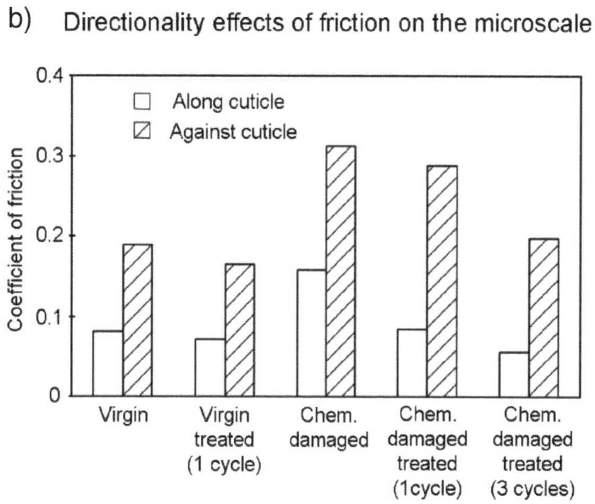

Fig. 5.26 (continued)

Coefficient of friction data showing directionality dependence on the macroscale for various virgin, damaged, and treated hair have been reported previously (Bhushan et al., 2005). Figure 5.26b shows a summary of the microscale coefficient of friction data for the various hair samples of this study. In most cases, the coefficient of friction has more than doubled when scanning in the against-cuticle direction. A more in-depth discussion on the coefficient of friction trends between the different hair types will follow. For now, however, it is important to realize that there is strong directionality dependence on coefficient of friction data for hair, especially on the microscale. On the nanoscale, while directionality dependence of friction force has been studied, actual coefficient of friction data are generally only measured on a small scan area which does not include the cuticle edge, so that only the true cuticle surface is involved (LaTorre and Bhushan, 2005a). Hence, nanoscale coefficient of friction directionality data similar to Fig. 5.26b are not shown.

5.3.2 Scale Effects on Coefficient of Friction and Adhesive Force of Various Hair

Given the fact that the "directionality effect" is now observed to be universal for all types of hair and on all scales, and since the along-cuticle direction is more relevant to our daily life (i.e., combing), we now focus on coefficient of friction data along the cuticle.

As described earlier, microscale and nanoscale coefficient of friction is taken as the slope of the least squares fit line of a friction force vs. normal force data curve. Figure 5.27a shows these types of raw data curves for various representative hair samples using both microscale (top plot) and nanoscale (bottom plot) AFM tips. The nanoscale data were taken from raw data presented by LaTorre and Bhushan (2006).

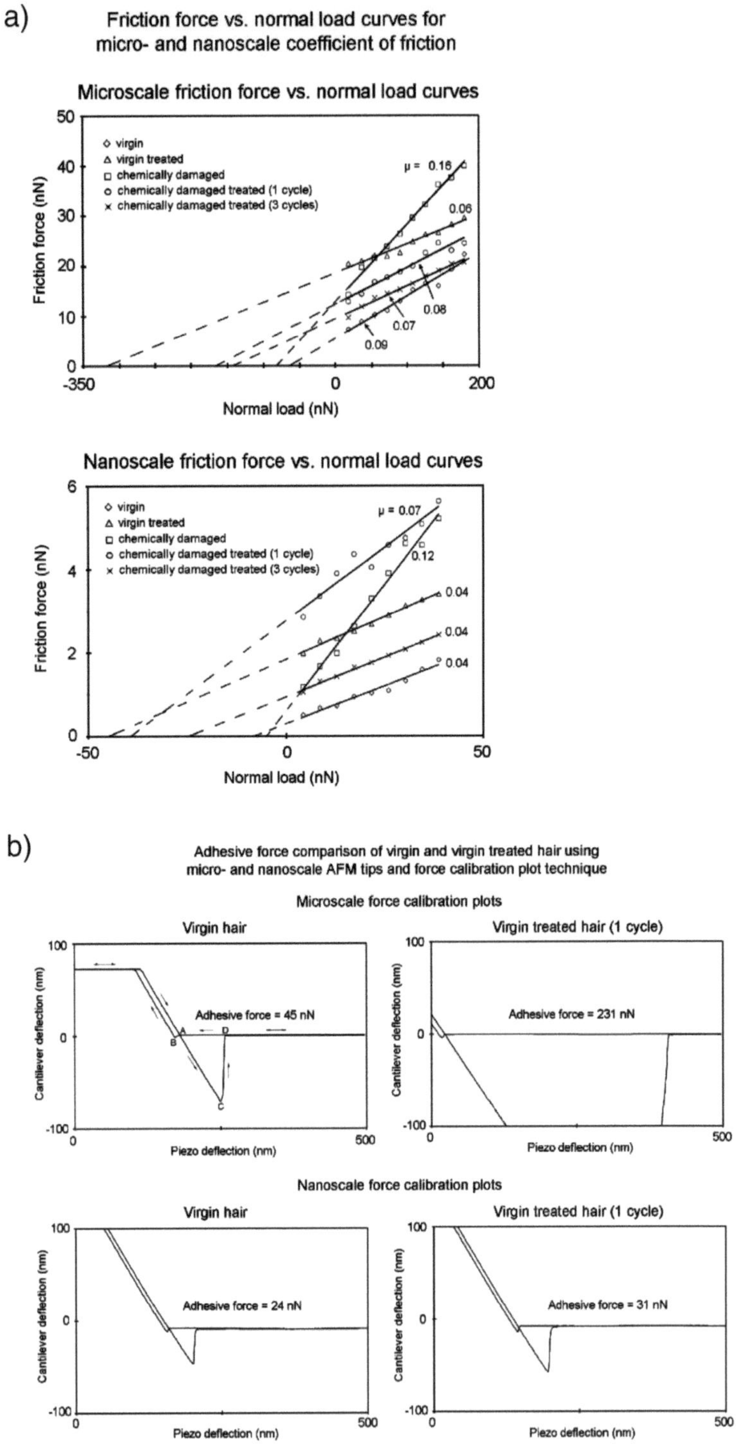

Fig. 5.27 (continued)

One can see a relatively linear relationship between the data points for each type of hair sample. If the least squares fit lines of Fig. 5.27a are extended to intercept the horizontal axis (indicated by the dotted lines), then a value for adhesive force can also be calculated using (5.1) These adhesive force values serve as values over the course of the full scan profile and differ slightly from the force calibration plots of Fig. 5.27b. From Fig. 5.27a, it is observed in both plots that treated hair fibers, whether virgin or chemically damaged, have much higher average adhesive force values as compared to their untreated counterparts. Chemically damaged hair is observed to have the highest coefficient of friction (highest slope) on both the micro- and nanoscales. As explained in LaTorre and Bhushan (2005a,b) and LaTorre and Bhushan (2006), chemical damage to the hair causes the outer lubricious layer of the cuticle to wear off, resulting in an increased coefficient of friction.

Force calibration plots yield adhesive force values at a single point and are considered to be more relevant for measurement of the pull-off force between the tip and hair surface. Since it has been previously shown that treating both virgin and chemically damaged hair with conditioner results in a large increase in adhesive force using both micro- and nanoscale AFM tips, we will focus only on the force calibration plots of the virgin and virgin treated hair to discuss mechanisms for this trend. The first plot in Fig. 5.27b shows a typical virgin hair force calibration plot with the microscale AFM tip. We can see that on the application of one cycle of conditioner to the virgin hair, the microscale adhesive force jumps to about 230 nN. This is clearly the effect of meniscus forces brought about by the presence of the conditioner layer on the cuticle surface which interacts with the tip (Chen and Bhushan, 2006; LaTorre and Bhushan, 2005a,b). With the nanoscale tip (bottom row of Fig. 5.22b), an increase in adhesive force is again seen with conditioner treatment.

It is important to notice that adhesive force values on the microscale are always larger than those on the nanoscale for a given hair. To explain the scale dependency of adhesive force, we can model the hair–conditioner–tip interaction as a sphere close to a surface with a continuous liquid film (Chen and Bhushan, 2006). The adhesive force $F_{a,}$ is the force needed to pull the sample away from the tip (which is the same as the adhesive force calculated with force calibration plots). As discussed earlier, F_a is the sum of van der Waals force F_{vdw} and the meniscus force F_m due to Laplace pressure ($F_a = F_{vdw} + F_m$), and F_m is given by (2.1). The increase in adhesive force calculated by force calibration plots with the microscale AFM tip compared to the nanoscale AFM tip is in large part due to the increased radius R of the microscale ball, which consequently induces larger F_m.

Fig. 5.27 (**a**) Friction force vs. normal force curves for micro- and nanoscale coefficient of friction of Caucasian virgin, chemically damaged, virgin treated, damaged treated (1 cycle conditioner), and damaged treated (3 cycles conditioner) hair; (**b**) adhesive force comparison of virgin and virgin treated hair using micro- and AFM tips and force calibration plot technique (LaTorre and Bhushan, 2006)

Figure 5.28 displays the coefficient of friction and adhesive force data on macro-, micro-, and nanoscales (LaTorre and Bhushan, 2006). Scale dependence is clearly observed. Macroscale data for virgin and virgin treated hair were taken from Bhushan et al. (2005). The values for other hair were taken in part from the indexed coefficient of friction values in LaTorre and Bhushan (2005b) which were transformed into actual values (as described previously). No adhesive force data is presented for the macroscale data because the adhesive force contribution to friction

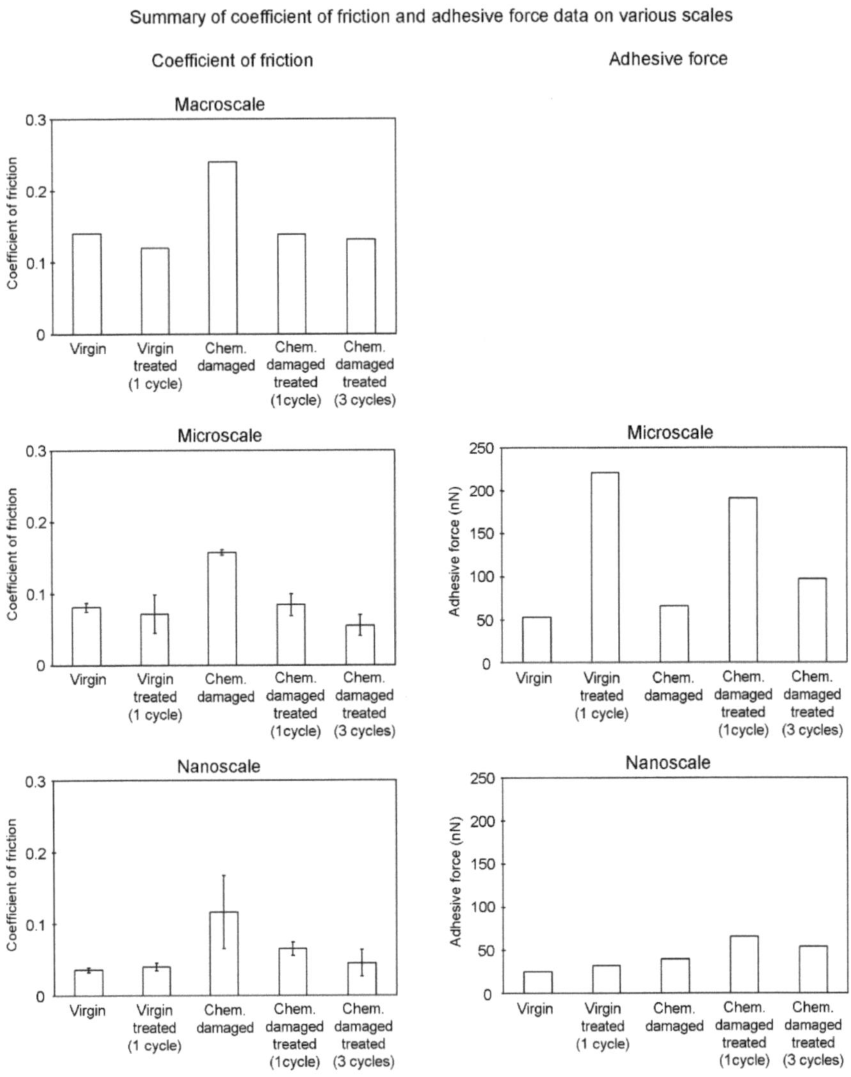

Fig. 5.28 Summary of coefficient of friction and adhesive force data on various scales for Caucasian virgin, chemically damaged, virgin treated, damaged treated (1 cycle conditioner), and damaged treated (3 cycles conditioner) hair (LaTorre and Bhushan, 2006)

is considered to be negligible compared to the applied normal load. On the micro- and nanoscales, however, the magnitude of the adhesive force is the same as that of the applied normal load, so they have significant contributions on the coefficient of friction data, and thus are presented. Microscale values are taken from the raw data represented in Fig. 5.28a and b. Nanoscale data were taken from LaTorre and Bhushan (2005b).

Macroscale coefficient of friction (COF) data are shown in the top row of Fig. 5.28. Chemically damaged hair has higher coefficient of friction than virgin hair, 0.24 compared to 0.14. Coefficient of friction decreases with the application of one cycle of conditioner for both virgin and chemically damaged hair. When three cycles of conditioner are applied to chemically damaged hair, there is only a slight decrease compared to one cycle application, 0.14 to 0.13. Thus, the data readily reveal that conditioner treatment decreases coefficient of friction for hair. The main mechanism for this macroscale trend is that lubrication with a thin conditioner layer occurs over a large contact area, and thus the conditioner layer shears easily to create a lubricious effect (LaTorre and Bhushan, 2005b). It is important to note that the magnitude of the coefficient of friction values is higher on the macroscale than on the other scales for all hair types. Bhushan et al. (2004) have previously outlined several differences in operating conditions which can be responsible for higher macroscale friction values. The one most relevant to our situation is that coefficient of friction increases with an increase in the AFM tip radius. Nanoscale friction data are taken with a sharp AFM tip, while the macro- and microscale tests have contacts which range from nanoasperities to much larger asperities which may be responsible for larger values of friction force on these scales. The combination of higher normal loads with a larger contact area (due to contact with multiple fibers at the same time) may also be responsible for increased coefficient of friction on the macroscale.

The coefficient of friction trends are similar on the microscale. On the microscale, virgin hair coefficient of friction was 0.08, while application of one cycle of conditioner decreased the coefficient of friction only slightly to 0.07. From Fig. 5.28 it is important to note the large standard deviation on the virgin treated value. The corresponding adhesive force data in the same row are useful to better understand this behavior. It is shown that the virgin treated hair has a large adhesive force contribution (due to meniscus effects caused by the conditioner layer) which has a significant effect on the variation of friction force, and consequently the coefficient of friction. The chemically damaged hair has the largest coefficient of friction of the set, 0.16. Application of one conditioner cycle brought the value down to 0.08, while it was even lower for three conditioner cycles, 0.06. Adhesive force increased significantly with conditioner application; for virgin hair, adhesive force increased from about 50 nN to 220 nN due to the meniscus effect that came about from AFM tip interaction with the conditioner on the cuticle surface. Likewise, the adhesive force for chemically damaged hair was about 65 nN and jumped to 190 nN and 100 nN for one and three cycles of conditioner treatment, respectively. A possible reason that one cycle of conditioner on damaged hair showed higher average adhesive force than three cycles could be that the three cycles generally places the conditioner more uniformly on the surface rather than accumulating it most on the bottom surface near the cuticle edge, which is where the adhesive force maps were generally taken. Nevertheless, the increased adhesive force shown in the plots is a clear indication of conditioner present on the hair surface (LaTorre and Bhushan, 2005b).

On the nanoscale, coefficient of friction of virgin and virgin treated hair is similar, but slightly higher for the treated cases. This is opposite of the trend on the macroscale and slightly different than that observed on the microscale. The meniscus bridges in the treated hair require more force to break through than with untreated hair, which causes the coefficient of friction to be similar to the untreated value, instead of experiencing a significant decrease which is traditionally expected. Damaged hair shows a much higher coefficient of friction and with more variation in the values since the chemical damage varies throughout each individual fiber. Contrary to the virgin and virgin treated hair nanoscale results, coefficient of friction for damaged hair decreased with application of conditioner treatment (both one and three cycles), which agrees well with macro- and microscale trends.

There are several reasons that nanoscale coefficient of friction trends for virgin treated and chemically damaged treated hair are different. The effect of chemical damage plays a large role. It is widely known that the cuticle surface of any hair is negatively charged. This charge becomes even more negative with the application of chemical damage to the hair. As a result, the positively charged particles of conditioner have even stronger attraction to the chemically damaged surface, which explains the increased presence of conditioner when compared to virgin treated hair. With the application of three conditioner cycles on damaged hair, there is an even more uniform placement of the conditioner. This leads to high adhesive force due to meniscus effect (similar to that of virgin treated hair) but, more importantly, lower shear strength on the surface. This creates an overall effect of lubrication as the tip travels across the cuticle surface and ultimately lowers the coefficient of friction.

Another reason for the difference in nanoscale trends between virgin and chemically damaged hair may have to do with drastic differences in hydrophobicity of the two hair types. Virgin hair has been shown to be hydrophobic, with a contact angle around 100° (Table 2.4). Chemically damaged hair, however, is hydrophilic, with a contact angle around 70°. The conditioner gel network is primarily composed of water, together with fatty alcohols, cationic surfactants, and silicones. Thus, the hydrophobicity of the hair will be relevant to not only how much conditioner is deposited, but also how it diffuses into the hair and bonds to the hair surface. For virgin treated hair, the conditioner deposits in certain locations, especially near the cuticle edge, but due to the hydrophobicity of the cuticle, it does not spread out as readily as with chemically damaged hair. For chemically damaged hair, the conditioner spreads out a bit more uniformly and in more places over the cuticle surface due to both the hydrophilicity and the stronger negative charge which attracts more conditioner deposition. Thus, as the tip scans over the virgin treated surface, the conditioner does not smear as readily, causing the tip to break the tiny meniscus bridges formed with the conditioner. This results in increased adhesive force, which contributes to higher friction force. This ends ups increasing the coefficient of friction to about the same level as the untreated virgin hair, instead of a reduction in coefficient which is typically expected for a lubricated surface. In the case of chemically damaged hair, however, the conditioner layer is already more spread out, especially in the case of three cycles of treatment. As the tip scans over the surface, the overall effect is one of reduced shear strength, i.e., the conditioner layer, albeit not fully continuous, smears with tip travel and causes reduced coefficient of friction between the tip and the hair. With three cycles, the conditioner thickness increases slightly, and the layer is even more uniformly distributed over the surface, which causes a further reduction in coefficient of friction, much like the results seen on both the micro- and macroscales.

Figure 5.29 shows schematically the mechanisms responsible for the reverse trends seen on the nanoscale for virgin treated hair (LaTorre and Bhushan, 2006). In the top cartoon of Fig. 5.29, the thin layer of conditioner acts as a lubricant over the

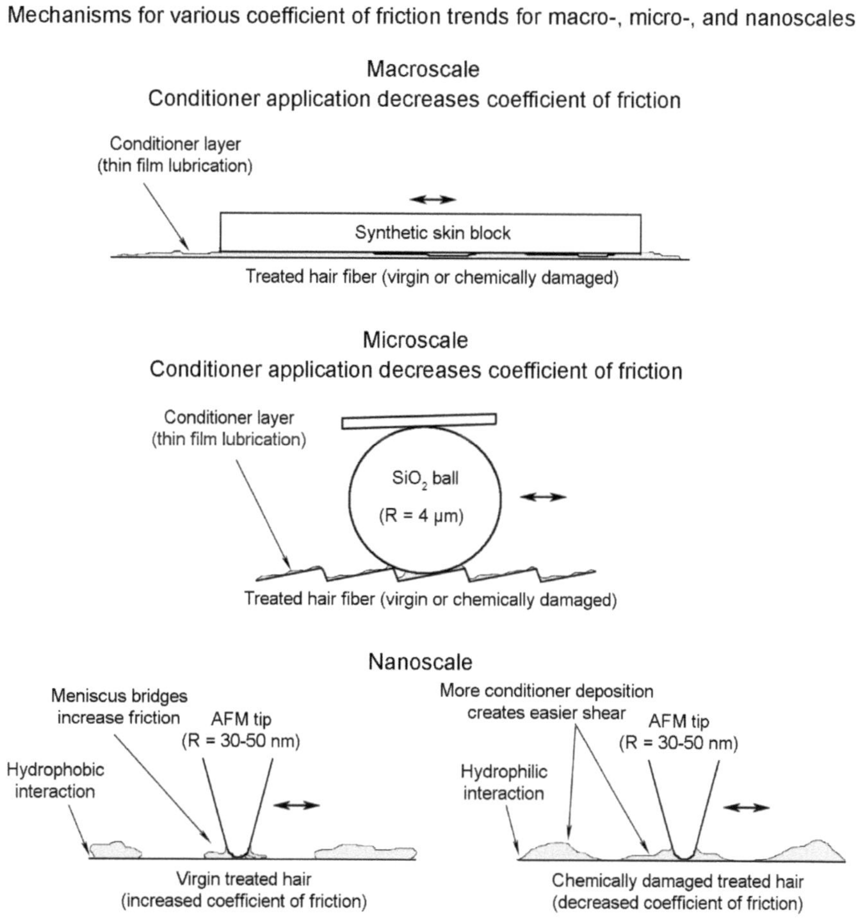

Fig. 5.29 Schematic of mechanisms for various coefficient of friction trends for macro-, micro-, and nanoscales (LaTorre and Bhushan, 2006)

hair fiber, limiting the dry contact with the synthetic skin block and creating easier relative motion, which decreases the coefficient of friction compared to the untreated hair. This is true on the macroscale for both virgin and chemically damaged hair. On the microscale, the same trend is experienced; that is, the 4 μm radius of the AFM ball comes in contact with multiple cuticle scales at the same time, causing an overall lubrication effect for both virgin treated and chemically damaged treated hair as the thin conditioner layer is sheared to create easier relative motion. It is important to note, however, that adhesive forces due to meniscus effects are of the same magnitude as the applied microscale normal load.

On the nanoscale (bottom cartoon of Fig. 5.29), we see different trends for virgin treated and chemically damaged treated hair. As discussed earlier, the hydrophobicity of the virgin hair causes different deposition of the conditioner. The AFM tip breaks the tiny meniscus bridges formed with the conditioner as it scans across the hair surface, which increases adhesive force contribution and results in an increased coefficient of friction. For hydrophilic chemically damaged hair, there is more uniform deposition and better smearing of the conditioner layer, which serves to lower coefficient of friction between the tip and the cuticle surface.

5.4 Summary

A flat-on-flat tribometer has been used to measure macroscale friction and wear of polyurethane film (synthetic skin) vs. hair and hair vs. hair. In the case of polyurethane film vs. hair, the chemo-mechanically damaged hair shows the highest coefficient of friction, followed by virgin and virgin treated hair. The coefficient of friction obtained in the case of hair vs. hair is greater than that of polyurethane film vs. hair. The coefficient of friction did not change 24 h after skin vs. hair wear test, while some of the cuticles were damaged.

AFM contact mode has been used to perform nanotribological studies on various hair and skin. Friction force and the resulting coefficient of friction are seen to be higher on chemo-mechanically damaged hair than on virgin hair, due to the increase in surface roughness and a change in surface properties that results from exposure to chemical damage. Generally speaking, the average coefficient of friction is similar between virgin and virgin treated hair of each ethnicity. However, in virgin treated hair there is an increase in friction forces around the cuticle edges and surrounding area. It is believed that the increase in friction force is due in part to an increase in meniscus effect, which increases the adhesive force contribution to friction at sites where conditioner is deposited or accumulated on the surface, namely around the cuticle scale edges. This adhesive force is of the same magnitude as the normal load, which makes the adhesive force contribution to friction rather significant. On the macroscale, however, the adhesive force is much lower in magnitude than the applied normal load, so the adhesive force contribution to friction is negligible over the hair swatch. As a result, treated hair shows a decrease in friction force on the macroscale, which is opposite of the micro/nanoscale trend. The friction and adhesion data on the micro/nanoscale are useful though, because they relate to the presence of conditioner on the cuticle surface and allow for obtaining an estimate of conditioner distribution. Studies using the force calibration plot technique showed a decrease in adhesive force with damaged hair, and significantly higher adhesive force for treated hair. This increase on the micro/nanoscale is most likely due to meniscus force contributions from the accumulation and localization of a conditioner layer on the hair surface. Thus, the presence of conditioner can be detected by this increasing adhesive force. The directionality dependence of friction is evident when the cuticle edge is examined using FFM.

Chemically damaged treated hair shows a much stronger affinity to conditioner than virgin hair. The negative charge of hair fibers becomes even more negative with the application of chemical damage to the hair. As a result, the positively charged particles of conditioner have an even stronger attraction to the chemically damaged surface, and this results in an increased presence of conditioner (and corresponding higher friction forces) when compared to virgin treated hair. With the application of three conditioner cycles on chemically damaged treated hair, friction force increases all over the cuticle surface, showing a more uniform placement of the conditioner. Contrary to the virgin and virgin treated hair results, coefficient of friction for chemically damaged hair decreased with application of commercial conditioner treatment (both one and three cycles). One possible explanation is that because the stronger negative charge on damaged hair results in better attraction of conditioner, this leads to higher adhesive force but, more importantly, lower shear strength on the surface.

Environmental effects were studied for various hair. The coefficient of friction generally decreased with increasing temperature. After soaking in de-ionized water, virgin hair exhibits a decrease in coefficient of friction after soaking. Virgin hair is more hydrophobic (based on contact angle data), so more of the water is present on the surface and results in a lubrication effect after soaking. Chemically damaged hair tends to be hydrophilic due to the chemical degradation of the cuticle surface and results in absorption of water after soaking. This softens the hair, which leads to higher friction, even with conditioner treatment. Durability tests show that once conditioner is applied to virgin hair, wear does not occur and there is no increase in friction force. Thus, conditioner serves as a protective covering to the virgin hair and helps protect the tribological properties when wear ensues.

In most cases, a decrease in coefficient of friction was observed on chemically damaged hair with the addition of the PDMS blend or amino silicones to the BTMAC surfactant. The silicones are typically used as a major source of lubrication and thus give the conditioner more mobility on the hair surface compared to just surfactants and fatty alcohols. The inverse trend was seen only for the amino silicone group. The dampened mobility of the amino silicone, with respect to hair surface and tip, may account for this wide variation in coefficient of friction. Adhesive force varied widely, but typically showed a significant increase with the presence of conditioner ingredients. This is a clear sign that meniscus effects are influencing the pull-off force between the tip and the sample. The amino silicones showed much more distinct regions of high and low friction and adhesion, which shows that there is less mobility of these molecules and much less redistribution as they coat the hair. Force calibration plots indicate that commercial conditioner containing only non-polar silicones and experimental conditioner containing polar amino silicones exhibit distinct affinity for chemically damaged hair surface. Commercial conditioner physically adsorbed on hair surface via van der Waals attractions can be easily squeezed out from the contact region under the load; on the other hand, amino silicones in experimental conditioner are chemically attached to the hair surface and cannot escape from the contact region under the load because of the strong electrostatic binding between the polar amino groups of silicone molecules and hair surface. Amino silicones provide much better load-bearing capacity and work as a cushion preventing hair surface from further damage.

Coefficient of friction values are largest on the macroscale, followed by microscale and then nanoscale values. In general, coefficient of friction increases with an increase in the AFM tip radius. Nanoscale friction data are taken with a sharp AFM tip, while the macro- and microscale tests have contacts which range from nanoasperities to much larger asperities which may be responsible for larger values of friction force on these scales. The combination of higher normal load with a larger contact area (due to contact with multiple fibers at the same time) may also be responsible for increased coefficient of friction on the macroscale as well.

On all scales, coefficient of friction decreases for chemically damaged treated hair compared to untreated. The same trend occurs for virgin treated hair on the macro- and microscales, but not on the nanoscale. On the larger scales, the thin layer of conditioner acts as a lubricant over the hair fiber across multiple cuticle contacts. On the nanoscale, the hydrophobicity of the virgin hair causes different deposition of the conditioner which increases the adhesive force contribution and results in an increased coefficient of friction. For chemically damaged hair, there is higher negative charge and a hydrophilic surface, which results in more uniform deposition and better smearing of the conditioner layer, which serves to lower coefficient of friction between the tip and the cuticle surface.

Chapter 6
Conditioner Thickness Distribution and Binding Interactions on Hair Surface

How common hair care products, such as conditioner, deposit onto and change hair properties are of interest in beauty care science, since these properties are closely tied to product performance. Conditioner is one of the hair care products which most people use on a daily basis. Conditioner thinly coats hair and can cause drastic changes in the surface properties of hair. Among all the components of conditioner, silicones are the main source of lubrication in the conditioner formulation. Silicone molecules remain as droplets surrounded by water, and their high molecular weight causes them to remain liquid and drain off hair surface gradually, which creates a long-lasting soft and smooth feel for conditioner-treated hair. The binding interaction between these molecules and hair surface is one of the important factors in determining the conditioner thickness distribution, and consequently the proper functions of conditioner.

The thickness and distribution of conditioner on the hair surfaces which affect the function of conditioner is of interest. In order to determine the thickness and distribution of thin liquid film on a substrate, several possible techniques can be applied: Fourier transform infrared spectroscopy (FTIR), ellipsometry, angle-resolved X-ray photon spectroscopy (XPS), and AFM (Bhushan, 1999b, 2008b). Ellipsometry and angle-resolved XPS have excellent vertical resolution on the order of 0.1 nm, but their lateral resolutions are on the order of 1 and 0.2 mm, respectively. Therefore, these techniques cannot be used for hair since the diameter of a hair fiber is only 50–100 μm. AFM on the other hand is a versatile tool and has a lateral resolution on the order of the tip radius (few nm), which is difficult to achieve by other techniques. In the study reported here, force calibration plot measurements using an AFM are conducted to obtain the local conditioner thickness distribution on various hair surfaces. The conditioner thickness is extracted by measuring the forces on the AFM tip as it approaches, contacts, and pushes through the conditioner layer. The conditioner thickness distribution on hair is effectively measured using an AFM. The effective Young's moduli of various hair surfaces are also calculated from the force distance curves using Hertz analysis. The binding interactions of different silicones on the hair surface, as well as the effect on their effective Young's modulus of the hair, are also discussed.

B. Bhushan, *Biophysics of Human Hair*, Biological and Medical Physics, Biomedical Engineering, DOI 10.1007/978-3-642-15901-5_6,

It has been reported earlier that, based on surface imaging using TR mode and friction force mapping on couple of cuticles, the conditioner is unevenly distributed across the hair surface, and thicker conditioner layer can be found near cuticle edges.

6.1 Conditioner Thickness and Adhesive Force Mapping

Figure 2.11, presented earlier, shows a typical force calibration plot for commercial conditioner treated hair, and the detailed interactions of the AFM tip, conditioner, and hair surface. Conceptually, the distance that the sample moves after the tip snaps in until it contacts the hair surface should be a measure of the thickness of the conditioner film. Here we define it as snap-in distance H_S. This snap-in distance H_s is not the real conditioner thickness h, and it tends to be thicker than the actual film thickness. In previous studies on determination of lubricant film thickness on a particulate disk surface by AFM (Bhushan and Blackman, 1991), it has been realized that as the ellipsometry thickness of the lubricant film increases, the thickness as measured by AFM also increases; however, AFM tends to measure a thicker film than measured by ellipsometry. There is a few nm offset between the snap-in distance H_s determined by AFM and film thickness h determined by ellipsometry. The most likely origin of the offset of the AFM thickness was attributed to a thin coating of the lubricant on the surface of the AFM tip that results from the tip being previously in contact with the lubricant film. This might account for part of the offset, but the main reason for this offset should be attributed to the deformation of the liquid film due to its interaction with the AFM tip as the liquid film approaches the tip (Chen and Bhushan, 2005, 2006; Lodge and Bhushan, 2006a). The liquid film can only approach the AFM tip to a finite minimal distance, below which the liquid surface is no longer stable due to the van der Waals attractive force between the liquid film and the AFM tip. For smaller distances, surface tension and adhesion to the substrate cannot keep the liquid surface from bulging and jumping into contact with the tip. Forcada et al. (1991) theoretically analyzed the hydrostatics of the liquid film in the force fields originated by the tip and solid substrate and indicated that the offset between the snap-in distance and the film thickness measured by ellipsometry arises from the bulging and posterior instability of the liquid film.

A number of theoretical publications (Attard and Miklavcic, 2001; Bhatt et al., 2001) have addressed the issue of liquid coalescence in terms of the "effective stiffness" of a liquid surface or interface and have concluded that a liquid surface behaves like a Hookean spring with an effective spring constant K_{eff} equal to its surface or interfacial tension γ, viz.,

$$K_{\text{eff}} \approx \gamma \tag{6.1}$$

Therefore, for one of the main components of conditioner, silicone (PDMS) in air, an effective spring constant of only 20 mN/m is expected, as the surface tension of PDMS is about 20 mN/m (Brandup et al., 1999) and water film has an effective spring constant of 72 mN/m. These are extremely low values compared to the cantilever spring constant of 5 N/m. It suggests that liquid surfaces can deform even by a very weak force. When the hair surface approaches the AFM tip, the weak van der Waals attractive force between the liquid conditioner film and the tip will cause the liquid film to deform and jump-in the tip at a finite distance. The theoretically expected van der Waals force F_{vdw} between a sphere and a flat surface is given by

$$F_{\text{vdw}} = HR/6D^2 \tag{6.2}$$

where H is the Hamaker constant which can be estimated based on the Lifshitz theory (Israelachvili, 1992), R is the radius of the AFM tip, and D is the distance between the AFM tip and the liquid conditioner film. Therefore, the jump-in distance D_J (the minimal distance between a stable liquid film and the AFM tip) can be theoretically calculated based on the criterion for a jump instability:

$$\mathrm{d}F_{\text{vdw}}/\mathrm{d}D = HR/3D_J^3 = K \tag{6.3}$$

where K is the spring constant. For conditioner-treated hair surface, K is the effective spring constant of the liquid film K_{eff} since it is much weaker than that of the cantilever. From (6.3), one obtains (Chen and Bhushan, 2006)

$$D_J = (HR/3K_{\text{eff}})^{1/3} \tag{6.4}$$

For a silicon tip covered by SiO_2 (the sphere) interacting with a liquid conditioner film (the flat surface) in air, the Hamaker constant H is estimated to be on the order of 10^{-20} J, R is about 100 nm, and K_{eff} is in the range of 20–72 mN/m (the surface tension of PDMS and water are 20 and 72 mN/m, respectively). Then a jump-in distance D_J of about 2 nm can be obtained based on (6.4). This jump-in distance is basically the offset between the snap-in distance H_s and the actual film thickness h. This 2 nm offset (jump-in distance) is surprisingly close to previous measurements on the localized lubricant film thickness on a particulate magnetic rigid disk, which indicated that the measured thickness using an AFM is about 2 nm larger than the actual thickness based on the ellipsometry measurements (Bhushan and Blackman, 1991). Note that the offset (jump-in distance D_J) only depends on the radius of the tip and the intrinsic properties of the film (surface tension and Hamaker constant), but is independent of the film thickness. Surface forces apparatus (SFA) experiments performed by Chen et al. (2004) indicated that for two 25 nm thick liquid PDMS films interacting in air at quasi-equilibrium state (very slow approach rate of about 0.3 nm/s), the jump-in distance is about 200 nm. SFA measurements gave much large jump-in distance because of the much larger value of radius. (In SFA experiments, the radius R is about 20 mm.) Therefore, although the snap-in distance H_s in force calibration plot overestimates the actual film thickness (approximately 2 nm thicker), it still provides a very good measurement for the actual thickness of thin conditioner film.

Figure 6.1 shows typical film thickness (snap-in distance H_s) and adhesive force mappings of virgin, chemically damaged, and chemically damaged treated (one

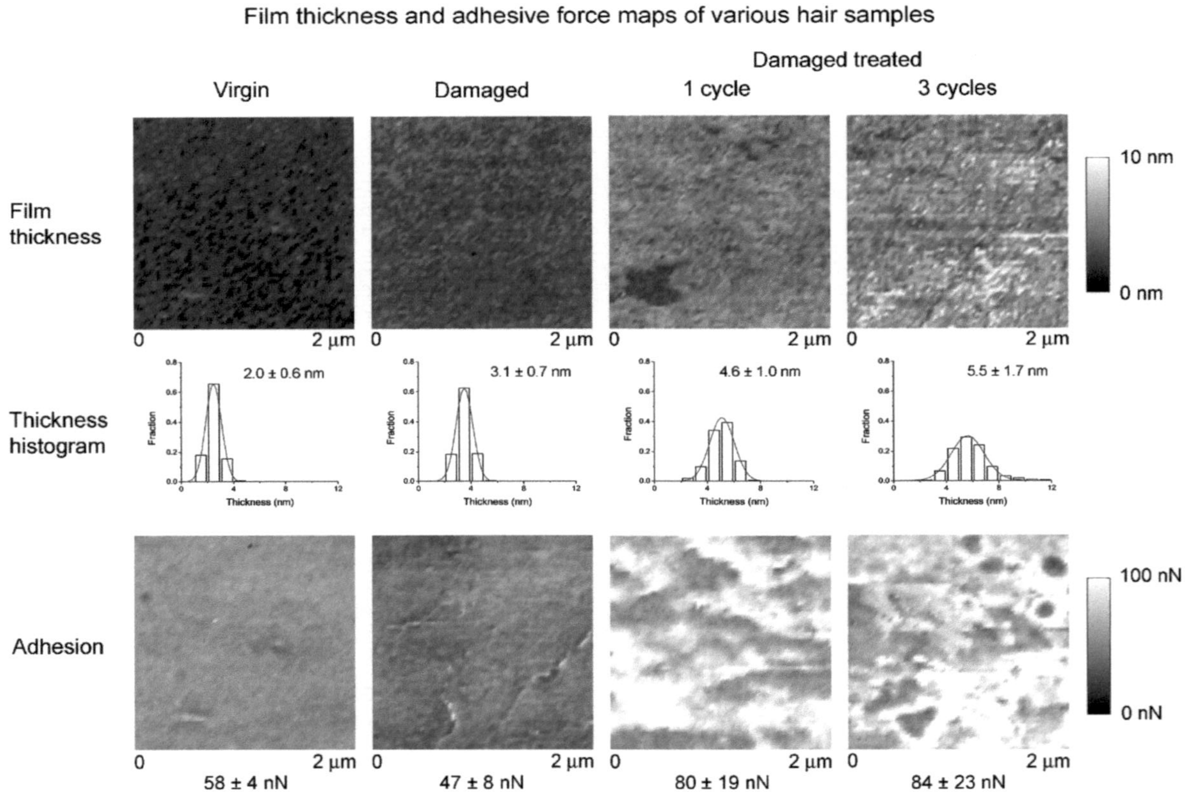

Fig. 6.1 Film thickness maps, histograms, and adhesive force maps of Caucasian virgin, chemically damaged, and damaged treated hair samples. The dotted lines in histograms are the Gaussian fits for film thickness (Bhushan and Chen, 2006)

cycle and three cycles) hair. The snap-in distance H_S of virgin hair is 2.0 ± 0.6 nm with a very narrow distribution, indicating that virgin hair surface is relatively uniform and not damaged. Virgin hair surface is covered with a saturated fatty acid lipid layer called 18-methyl eicosanoic acid (18-MEA) (Robbins, 1994), which makes the hair surface hydrophobic (see Table 2.4 for contact angle data). Therefore, virgin hair surface does not consist of continuous water film. Instead of the deformation of the liquid film, the AFM tip will jump into contact with the hair surface at a finite tip–sample distance because of the van der Waals attractive force.

The snap-in distance of chemically damaged hair is larger than that of virgin hair and increased to 3.1 ± 0.7 nm. The outermost surface of hair is the cuticle which consists of large amounts of cystine. Chemical treatment will partially remove the fatty acid lipid layer (18-MEA) covering the hair surface and break the disulfide bonds in cystine to form new ionic groups (such as cysteic acid residues by oxidation of cystine acids). The hair surface becomes hydrophilic, and the contact angle with water decreases (Molina et al., 2001). Consequently, the amount of water adsorbed on the hair surface increases. As the sample approaches the tip, the AFM tip will jump into contact with the hair surface at a finite distance because of van der Waals attractive force, as well as small meniscus bridge formation between the tip and adsorbed water layer. The adhesive force of chemically damaged hair surface is 47 ± 8 nN, less than that of virgin hair (58 ± 4 nN). The existence of a thin adsorbed water layer tends to decrease van der Waals attractive force but increases the meniscus force. Note that the tip and the hair surface are not as smooth as shown in Fig. 2.10 (rough on the molecular scale), and the adsorbed water layer is very thin; therefore, no single large meniscus but many small menisci form between the tip and the hair surface. The total attractive force of chemically damaged hair therefore is smaller than that of virgin hair.

After conditioner treatment, the snap-in distance H_S increases with the number of cycle of treatment. The snap-in distances H_S for one cycle treatment and three cycles treatments are 4.6 ± 1.0 and 5.5 ± 1.7 nm, respectively. Thicker liquid film is unevenly present on conditioner-treated hair surface, and the thickness tends to have a broader distribution than that of chemically damaged hair. It is more obvious for three cycles damaged treated hair, in which the thickness distribution is much broader and tends to have a long tail to larger thickness. The measured mean thickness varies from 5 up to 25 nm (data not shown). Excluding the 2 nm offset due to the jump-in deformation of the liquid conditioner film, the actual film thickness should be around 3–23 nm. These values are consistent with the previous estimated conditioner thickness based on the amount of material deposition (LaTorre et al., 2006) (see Appendix B). The adhesive forces for one cycle and three cycles are 80 ± 19 and 84 ± 23 nN, respectively. The larger adhesive force is attributed to the formation of large meniscus between the tip and the conditioner layer on the hair surface.

6.1.1 Effect of Humidity and Temperature on Film Thickness and Adhesion

Figure 6.2 shows typical film thickness mapping, histogram of film thickness, and adhesive force mappings of damaged treated hair at different humidities (Bhushan and Chen, 2006). At different humidities, the film thickness and adhesive force of hair are distinctly different. At low humidity, the treated hair surface has only a very thin layer of film (2.6 $\pm$ 0.7 nm). At high humidity, a much thicker film is present on the hair surface (4.7 $\pm$ 1.8 nm). This film tends to extend a much longer tail into the thicker film regime. At high humidity, hair surface has a higher adhesive force compared to that at low humidity. The conditioner-treated hair surface will be covered with a gel network. This gel network can adsorb or desorb water depending on the environmental condition. The conditioner-treated hair surface tends to adsorb more water as the humidity increases, which will increase the film thickness on hair surface and consequently increases the adhesive force.

Figure 6.3 summarizes the conditioner thickness and adhesive force of various hair surfaces at different humidities (Bhushan and Chen, 2006). Film thickness on virgin hair remains relatively constant at different humidities, and it only increases slightly at high humidity. The outer layer of the hair is covered with hydrophobic 18-MEA and remains intact in virgin hair. Therefore, water is hardly adsorbed (or desorbed) and penetrated into hair surface, and humidity has little effect on film thickness of virgin hair. On the other hand, the hydrophobic lipid layer of damaged hair surface may be depleted or damaged; therefore, inner cellular structures of hair, which consist of many hydrophilic molecules, are now exposed to water. Damaged hair surface is partially hydrophilic and will be more sensitive to the humidity. It tends to adsorb more water at high humidity; therefore, film thickness of damaged hair increases gradually as the humidity increases. Damaged treated hair is covered with a conditioner film; therefore, it has a thicker film thickness. Film thicknesses of treated hair surface typically show large deviations compared to those of virgin and damaged hair, indicating uneven distribution and adsorption of conditioner layer on hair surface. The film thickness increases as the humidity increases because the conditioner gel network layer can easily adsorb more water at high humidity. However, this conditioner gel network layer fails to retain water content at low humidity. As shown in Fig. 6.3, virgin hair usually has a higher adhesive force than damaged hair. Adhesive force includes the van der Waals attraction and meniscus forces. Virgin hair has larger van der Waals force than damaged hair because of the hydrophobic nature of the virgin hair surface. Treated hair has higher adhesive force due to conditioner meniscus formation.

The film thickness and adhesive force were also measured at different temperatures (Bhushan and Chen, 2006). Figure 6.4 summarizes the data for virgin, damaged, and conditioner-treated hair at those temperatures. Temperature has little effect on the film thickness of virgin or damaged hair in the temperature range studied. Conditioner-treated hair has thicker film compared to virgin or damaged hair. As temperature increases, the thickness of conditioner layer starts to decrease.

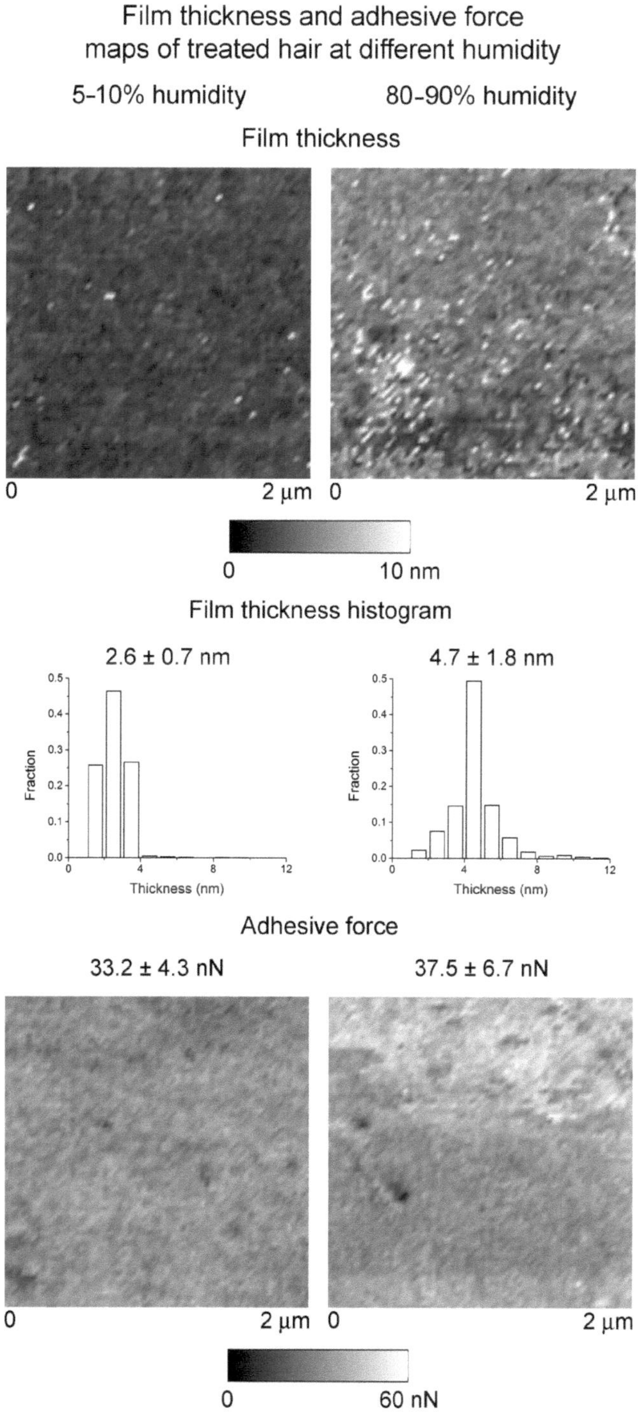

Fig. 6.2 Typical film thickness and adhesive force maps of damaged treated hair at different humidities (Bhushan and Chen, 2006)

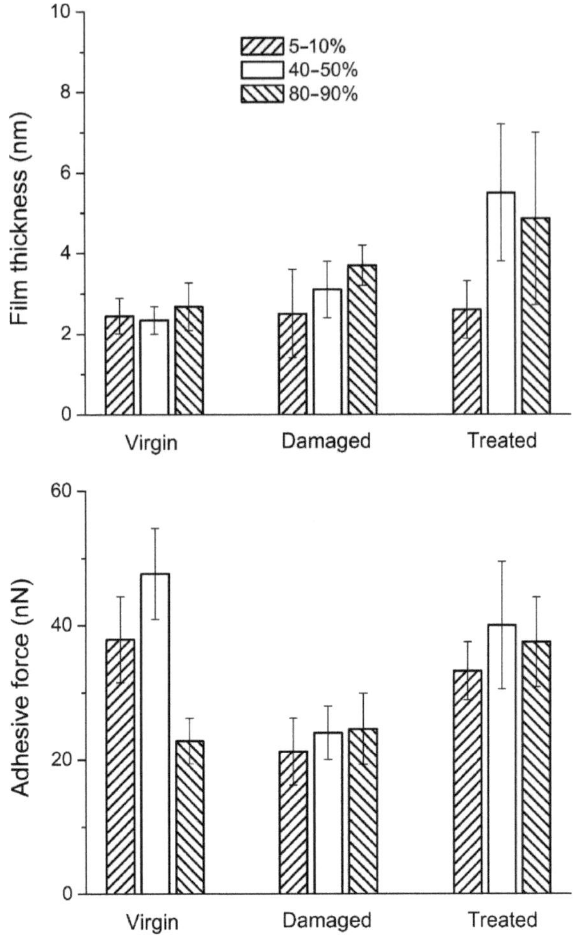

Fig. 6.3 Film thickness and adhesive forces of Caucasian virgin, chemically damaged, and damaged treated hair samples at different humidities. Large deviations compare to average values, especially for conditioner-treated hair surface, indicating uneven distribution and adsorption of conditioner layers (Bhushan and Chen, 2006)

At high temperature, conditioner layer will eventually lose all water content. In the studied temperature range, there is little effect on the adhesive force of various hair surfaces.

6.2 Effective Young's Modulus Mapping

The adhesive force as well as the total force acting on the tip at each measurement point can be accurately measured from the force calibration plot. If the zero tip–sample separation is defined to be the position where the force on the tip is zero when in contact with the sample, the force calibration plot (cantilever deflection vs. piezo position plot as shown in Fig. 2.11) can be converted to a force vs. tip–sample separation curve (Chen and Bhushan, 2006). Figure 6.5a shows the forces acting on the tip as a function of tip–sample separation for virgin, commercial conditioner

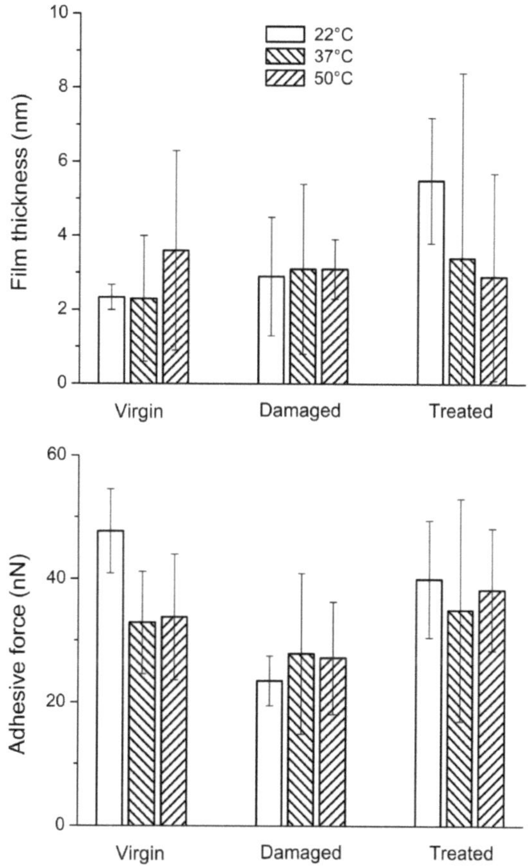

Fig. 6.4 Film thickness and adhesive forces of Caucasian virgin, chemically damaged, and damaged treated hair samples at different temperatures (Bhushan and Chen, 2006)

(PDMS blend silicone) treated, and experimental conditioner (with amino silicone) treated hair. The lowest point on the approach curve is assumed to be the point that the tip contacts the hair surface. Afterward, the hair surface deforms elastically under the load, from which the deformation (indentation Δz) of the surface can be extracted. Plotting the obtained deformation (indentation Δz) against the total force (on approach curve) acting on the surface gives force vs. indentation (deformation) curves. Figure 6.5b shows the force vs. indentation curves for virgin, commercial conditioner treated, and experimental conditioner (amino silicone) treated hair surfaces which are extracted from the force vs. tip–sample separation curves shown in Fig. 6.5a, and the effective Young's modulus of various hair surfaces can be determined from these curves by fitting them to (2.2). The effective Young's modulus of virgin, commercial conditioner treated, and experimental conditioner treated hair surfaces are 5.3 ± 0.9, 0.60 ± 0.03, and 0.032 ± 0.002 GPa, respectively, for these three specific curves. For these calculations, the radius of the tip is measured to be approximately 100 nm. The effective Young's modulus of a conditioner-treated hair surface can be one to two orders of magnitude less than that of virgin hair.

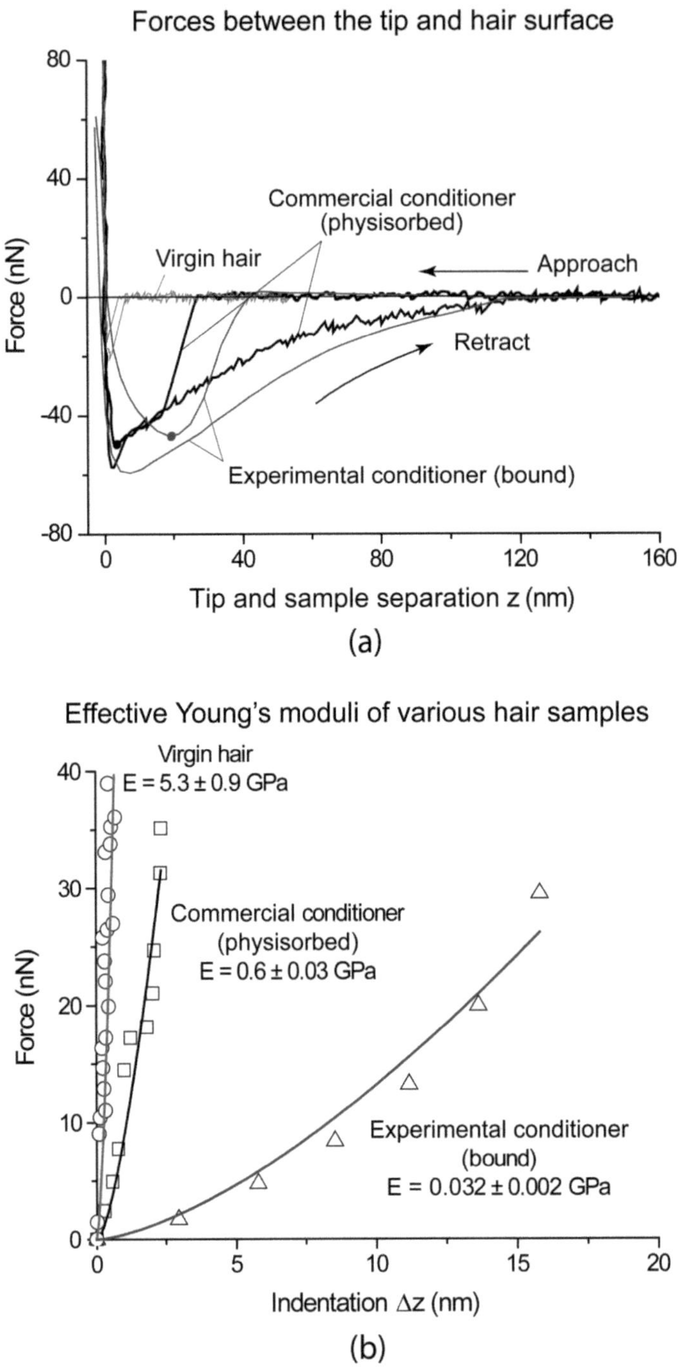
Forces between the tip and hair surface
Force (nN)
80
40
0
-40
-80
Commercial conditioner
(physisorbed)
Virgin hair
Approach
Retract
Experimental conditioner (bound)
0
40
80
120
160
Tip and sample separation z (nm)
(a)
Effective Young's moduli of various hair samples
Virgin hair
E = 5.3 ± 0.9 GPa
Commercial conditioner
(physisorbed)
E = 0.6 ± 0.03 GPa
Experimental conditioner
(bound)
E = 0.032 ± 0.002 GPa
Force (nN)
40
30
20
10
0
0
5
10
15
20
Indentation Δz (nm)
(b)

Repeating these calculations over the whole surface on various hair samples, the maps of the effective Young's modulus of various hair surfaces are obtained as shown in Fig. 6.6. Virgin hair surface has an effective Young's modulus of about 7.1 ± 2.9 GPa, which is the most stiff among all the samples and is consistent with previous nanoindentation measurement results (Wei et al., 2005). The disulfide-rich cross-links on the outermost layer of the hair surface account for this stiffness; the effective Young's modulus of chemically damaged hair surface (6.6 ± 3.3 GPa) is slightly smaller than that of virgin hair surface. The chemical treatment partially breaks the disulfide cross-links in the outermost layer of the hair surface and weakens the hair surface. Commercial conditioner treated hair surface tends to have much smaller effective Young's modulus than that of chemically damaged hair surface (5.5 ± 4.6 GPa for one cycle treatment, and 3.3 ± 3.2 GPa for three cycles treatment). The effective Young's modulus decreases as the number of cycles of treatment increases. Large deviations of the effective Young's modulus of conditioner-treated hair surfaces are seen, indicating the uneven distribution and adsorption of conditioner on hair surface. This large deviation of the effective Young's modulus of conditioner-treated hair surface is due to a large area of the hair surface that has a low effective Young's modulus (dark region). Comparing the conditioner thickness map (see Fig. 6.1) and the effective Young's modulus map (see Fig. 6.6) of three cycles treated sample, it is clear that the region with small effective Young's modulus (dark region) is closely correlated to the region with thick conditioner layer (bright region) on the hair surface. The effective Young's modulus decreases as conditioner thickness increases. Therefore, although the average value of the effective Young's modulus of three cycles treated hair surface is 3.3 GPa, some part of the surface can have very low effective Young's modulus value (as low as the 0.60 GPa value shown in Fig. 6.5b). Experimental conditioner (amino silicone) treated hair surface has extremely low effective Young's modulus (0.047 ± 0.045 GPa) compared to the value of chemically damaged hair surface. The film thickness, adhesive force, and effective Young's modulus of various hair samples are summarized in Table 6.1 (Chen and Bhushan, 2006).

6.3 Binding Interactions Between Conditioner and Hair Surface

Small molecules in conditioner, such as water and surfactant, may diffuse into the outermost layers of the hair surface and swell the disulfide cross-link network within these layers, which in a way weakens these layers. This consequently reduces the

Fig. 6.5 a The force acting on the tip as a function of tip–sample separation for various hair samples. The sample is moved with a velocity of 200–400 nm/s, and the zero tip–sample separation is defined to be the position where the force on the tip is zero when in contact with the sample. The *arrows* show the direction of motion of the sample relative to the tip. A negative force indicates an attractive force. (**b**) Surface-effective Young's modulus extracted from the force distance curves. The *solid lines* show the best fits for the data points of various hair samples (Chen and Bhushan, 2006)

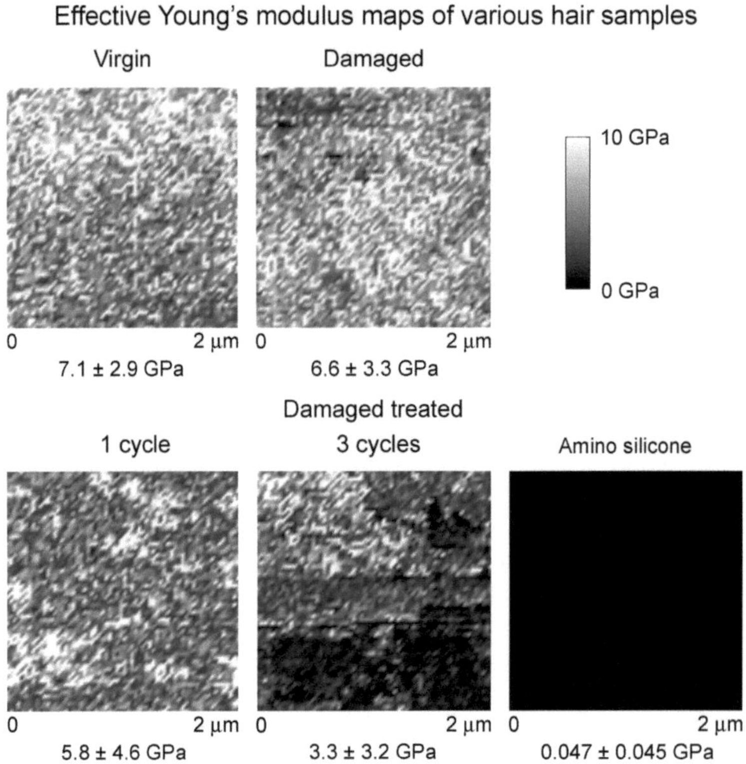

Fig. 6.6 Effective Young's modulus map of Caucasian virgin, chemically damaged, and damaged treated hair samples. The average value and standard deviation for each map are listed. Large deviations compare to average values, especially for conditioner-treated hair surface, indicating uneven distribution and adsorption of conditioner layers (Chen and Bhushan, 2006)

Table 6.1 Summary of film thickness, adhesive force, and effective Young's modulus of various Caucasian hair samples

Samples		Film thickness (nm)	Adhesive force (nN)	Young's modulus (GPa)
Virgin		2.0 ± 0.6	58 ± 4	7.1 ± 2.9
Chemically damaged		3.1 ± 0.7	47 ± 8	6.6 ± 3.1
Chemically damaged, treated	1 cycle	4.9 ± 1.0	80 ± 19	5.8 ± 4.6
	3 cycles	5.5 ± 1.7	84 ± 23	3.3 ± 3.2
	Amino silicone	n/a	n/a	0.047 ± 0.046

effective Young's modulus for conditioner-treated hair surfaces. Even more significantly, although silicones in conditioner cannot diffuse into the hair surface layers because of the size of the molecules, they can physically adsorb via van der Waals interaction or bind (chemically or electrostatically) on the hair surface, which will significantly affect the physical properties of the hair surface. Although the AFM cannot give an absolute distance as the surface forces apparatus (SFA)

(Israelachvili, 1992), the force distance curves obtained using an AFM still can show the interactions between the surface and the tip, and consequently reveal the essence of various surfaces. As shown in Figs. 2.11 and 6.5a, virgin hair and conditioner-treated hair behave differently as the tip approaches, compresses, and retracts from the surface. Virgin hair surface behaves like a stiff elastic solid surface. As the virgin hair surface approaches the tip, the tip will jump into contact with the hair surface at a small separation of around 4 nm because of the van der Waals attractive interaction between the hair surface and the tip. Then the force quickly goes to repulsive and the tip reaches the hard wall contact (tip–hair solid–solid contact) with very small deformation. When the virgin hair surface is withdrawn, the tip will simply jump out, and no large liquid deformation occurs because of the lack of liquid layer.

Conditioner-treated hair surfaces show a much longer range of interactions than virgin hair because of the presence of liquid conditioner layer. The tip will snap in at a large separation because of the van der Waals force as well as meniscus formation between the tip and conditioner liquid layer. When the sample is pulled away from the tip, the forces on the tip will slowly decrease to zero as a long meniscus bridge of liquid is drawn out from the surface and eventually breaks at a large separation. Another important feature is that before the tip reaches the tip–hair solid–solid contact, commercial conditioner treated hair surface shows larger deformation than virgin hair surface, indicating the presence of a thin physically adsorbed layer on the hair surface. Experimental conditioner treated hair surface shows very large deformation, and no tip–hair solid–solid contact is reached under the given experimental conditions, indicating there is a strongly bound conditioner layer. Large hysteresis between approaching and retracting curves indicates that this bound layer deforms viscoelastically under the load. The adhesive force (on the approach curve) of experimental conditioner treated hair surface is smaller than that of commercial conditioner treated hair surface. The meniscus forces in both cases would be similar, and the extra repulsive force on the tip for experimental conditioner treated surface comes from the compression of the molecules underneath the tip. The tip has to overcome the intermolecular polar or hydrogen bonding and strong binding between amino silicones and the hair surface so as to squeeze out silicone molecules from between the tip and the hair surface.

Figure 6.7 illustrates the difference between physisorbed conditioner and bound conditioner. Chemical treatment breaks the disulfide bonds of the cystine cross-link network in the cuticle and forms new ionic groups on the hair surface; therefore, chemically damaged hair surface is negatively charged. Silicones in commercial conditioner are non-polar and do not contain any functional groups; therefore, most of the silicone molecules are free and mobile except the last layer of molecules adjacent to the hair surface which may be physically adsorbed via van der Waals interactions. As the tip starts to touch and compress the conditioner film, free conditioner molecules can easily escape from the contact region. The last layer of adsorbed molecules may sustain some load before they are squeezed out eventually at higher load. Then the tip penetrates the conditioner layer and reaches the tip–hair solid–solid contact and compresses directly on the hair surface. The last layer of

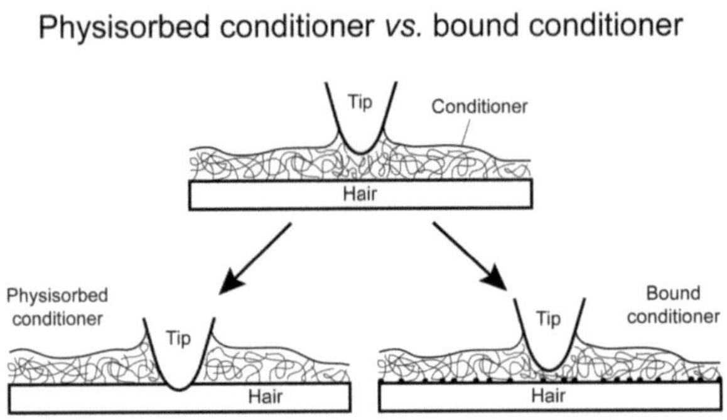

Fig. 6.7 Comparison between physisorbed conditioner and bound conditioner. Physisorbed conditioner can be easily squeezed out from the contact region under the load; on the other hand, bound conditioner cannot escape from the contact region under the load because of the strong electrostatic binding between the silicone molecules and the hair surface (*black dots* are the binding sites) (Chen and Bhushan, 2006)

physisorbed conditioner molecules is very thin but effectively lowers the Young's modulus of the hair surface by one order of magnitude (see Figs. 6.5b and 6.6).

On the other hand, the experimental conditioner consists of amino silicones, which are positively charged in an aqueous condition and can strongly bind with the anionic groups present on the chemically damaged hair surface. The binding between amino silicone molecules and the hair surface is much stronger than van der Waals attractions in commercial conditioner treated hair surface, which makes bound conditioner molecules difficult to escape from the contact region even at high load. Therefore, after the tip touches the conditioner, instead of penetrating the conditioner layer and compressing directly on the hair surface, it compresses on a soft bound conditioner layer. The effective Young's modulus of this layer (silicones) is about 0.047 ± 0.045 GPa. Although it is much softer than hair, it is still much stiffer than bulk PDMS whose Young's modulus is only 680 kPa (Mark, 1999). Two reasons are attributed to this increase of effective Young's modulus of silicones (PDMS): the bound silicone layer is very thin (a few nm) and it is in a confined geometry during the measurements. It is well known that thin films can exhibit different physical properties compared to bulk materials. Thin liquid film can behave like a solid under confinement (Israelachvili, 1992). On the other hand, underneath this thin bound silicones layer is the stiff hair surface whose Young's modulus is much higher (chemically damaged hair surface 6.6 ± 3.3 GPa, see Fig. 6.6). Since the bound layer is very thin, the measured Young's modulus therefore has the contributions from both the soft bound silicone layer and the stiff hair surface.

Amino silicones work as a cushion protecting the hair surface. Although this strong affinity to the hair surface can dramatically increase the load or contact pressure that a conditioner film can support before solid–solid contact, an AFM study on hair (LaTorre et al., 2006) indicated that treated hair with excess amino silicone has higher coefficient of friction than commercial conditioner (physisorbed) treated hair. It is well known that in general, lubricant film with mobile and immobile fractions (partially bonded lubricant film) is desirable for low friction and high durability (Bhushan, 1999a, 2002). Biolubrication studies using SFA (Benz et al., 2004) also indicated that a low coefficient of friction is not necessarily a measure of good wear protection. A conditioner layer strongly bound on the hair surface is good for preventing hair from further damage, but it may not be good from the lubrication point of view. A balance between good lubrication and good protection has to be reached for a good conditioner product.

6.4 Summary

The snap-in distance H_S in the force calibration measurement provides a good estimate for the conditioner film thickness on hair. The conditioner unevenly distributes on hair surface. Conditioner thickness increases with the number of cycles of treatment. The mean conditioner thickness is on the order of 5 nm. The adhesive force of damaged hair was smaller than that of virgin hair because of different surface hydrophobicities. The treated hair showed a much higher adhesive force because of meniscus formation at the interface. Humidity had little effect on film thickness and adhesive force for virgin hair; however, it had a larger effect on damaged and treated hair. Temperature effect was little on all hair. The effective Young's modulus of virgin hair surface was the highest, damaged hair was slightly smaller, and treated hairs had the smallest. Amino silicone conditioner showed strong affinity to hair and strengthened the hair fiber and is expected to work as a cushion for protecting the hair surface from further damage.

Chapter 7
Surface Potential Studies of Human Hair Using Kelvin Probe Microscopy

It is obvious that during combing and running the hands through one's hair, physical damage is likely to occur. These actions also tend to create an electrostatic charge on the hair, and it is of interest whether or not the physical wear and the electrostatic charge are related, or if charge has been created by other mechanism. The presence of this static charge in the real world is a major problem concerning hair manageability, flyaway behavior, feel, and appearance so understanding the mechanisms behind charge buildup, and how to control it, is a focal point in designing effective hair care products such as conditioner. Conditioner is known to affect the surface potential characteristics of hair, and understanding this behavior on the small scale is of great interest. Kelvin probe has been used to provide great insight into the mechanisms behind the electrostatic behavior of hair. Electrostatic charging of hair has been studied by charging the hair with latex, a material that is known to create a static charge on hair and by applying an external voltage. Hairs with varying types and degrees of damage as well as varying types and degrees of conditioner treatments have been studied. In a macroscale rubbing process, it is difficult to control the rubbing load and rubbing area. Hence, for systematically studying charging characteristics, it is essential to charge hair surface on a nanoscale by a controlled process as well.

7.1 Effect of Physical Wear and Rubbing with Latex on Surface Potential

Lodge and Bhushan (2007a) measured surface potential change after performing wear experiments in an AFM on hair fiber using a diamond tip with a tip radius of 60 nm at a load of 5 μN. The wear scars were created at 2 μm $\times$ 2 μm size for one pass. After samples were worn with the diamond tip, the AFM tip was changed to a conductive tip, and the surface potential was mapped in the wear area. All samples were mounted on a sample puck in conductive silver paint. They found that physical wear alone does not cause a measurable electrostatic charge on hair because hair is non-conductive.

The electrostatic charge buildup on hair was caused by rubbing with a natural latex finger cot, and surface potential change was measured after rubbing (Lodge

B. Bhushan, *Biophysics of Human Hair*, Biological and Medical Physics, Biomedical Engineering, DOI 10.1007/978-3-642-15901-5_7,

and Bhushan, 2007a) First, a surface potential map of the hair sample, which was electrically isolated from ground to prevent quick discharging, was obtained for "before rubbing" data. Then samples were wiped lightly with the finger cot five times along the hair shaft in the direction from root to tip. Surface potential was then mapped immediately for "after rubbing" data. Figure 7.1a and b shows AFM images for virgin, virgin treated, chemically damaged, chemically damaged treated (one cycle), and chemically damaged treated (three cycles) hair samples. These images are presented in three columns. The first is the surface height of the sample, and the second is the absolute surface potential. The final column shows the same surface potential data as the second column, but the average surface potential (which is presented in the subsequent bar charts) is subtracted out, and the scale is reduced to show more contrast. Before rubbing, data were taken as a baseline in order to report a change in potential due to rubbing. Absolute quantitative values of surface

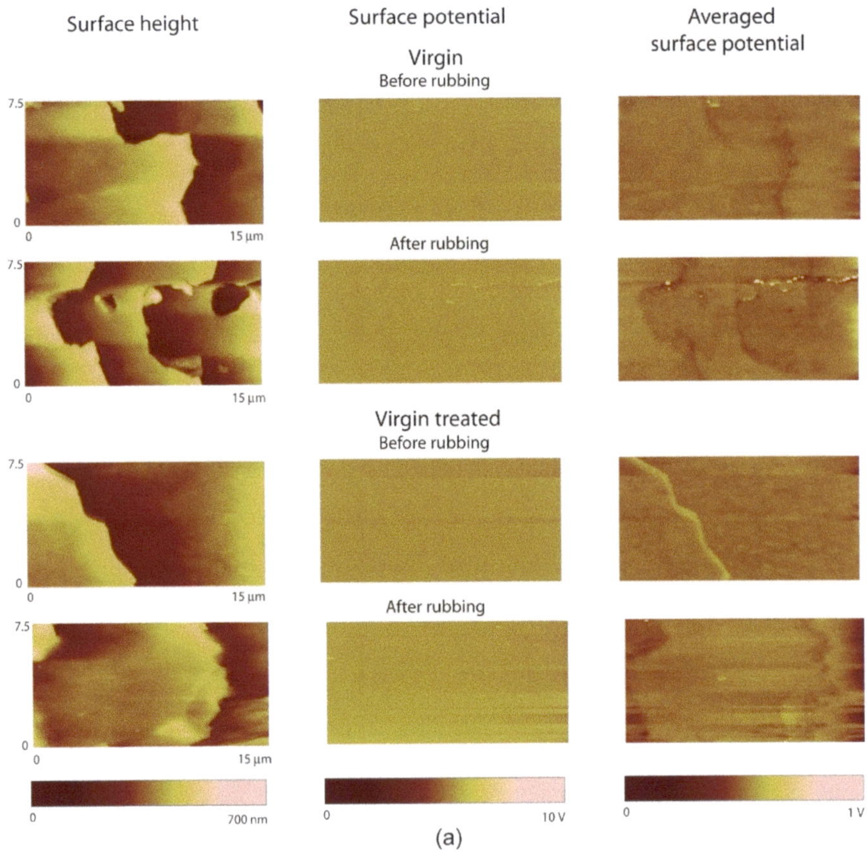

Fig. 7.1 (**a**) AFM images of virgin and virgin treated hair samples both before and after rubbing with latex; (**b**) chemically damaged, chemically damaged treated (1 cycle), and chemically damaged treated (3 cycles) hair samples both before and after rubbing with latex; (**c**) bar charts of average surface potential change after rubbing with latex (Lodge and Bhushan, 2007a)

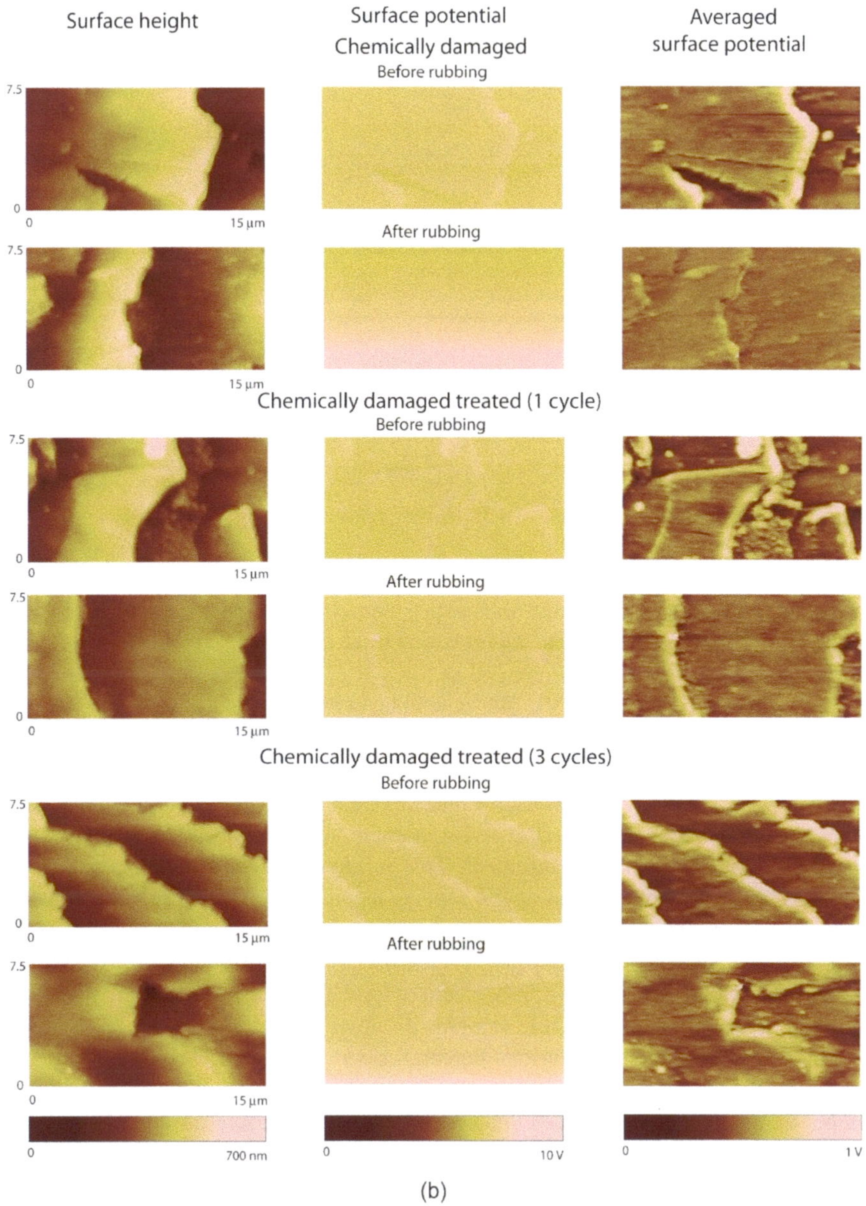

Fig. 7.1 (continued)

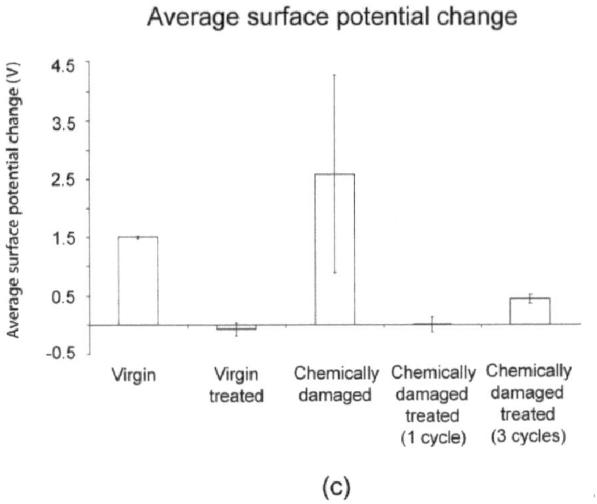

Fig. 7.1 (continued)

potential were not reliable for this experiment because the samples were not grounded, thus the necessity to report a potential change. It should be noted that some samples became highly charged as a result of the rubbing, beyond the capability of the microscope (10 V). These samples were not used in this study. However, it is important to note that all of the samples tested are capable of developing high amounts of charge. A method where the force of contact is more accurately controlled may provide better insight into this behavior. Some samples, especially the chemically damaged one (Fig. 7.1a), show a significant amount of change in surface potential during measurement. This is illustrated as a higher potential at the bottom of the image than at the top, as the measurement was started at the bottom. This indicates that the charge dissipates relatively rapidly and seems to reach a constant value by the time the measurement is finished. The fact that the virgin and treated samples (Fig. 7.1a) seem to have a more constant potential during measurement indicates that these samples dissipate charge faster, and by the time the measurement is taken a constant potential is more or less achieved. The average surface potential change is shown in Fig. 7.1c. This bar chart indicates that conditioner treatment greatly reduces the amount of charge present on the hair surface. It also shows that chemical damage increases the amount of charge built on the hair surface.

Figure 7.2 shows chemically damaged hair that has been treated with an experimental conditioner containing amino silicone, which is believed to chemically attach to the hair surface. Very little difference is seen between the amino silicone conditioner and the commercial PDMS silicone conditioner. (Commercial PDMS silicone treated hair data are shown in Fig. 7.1b). This is likely because the latex rubbing occurs over a very large area of the hair, and any difference in conditioner distribution is occurring on a much smaller scale. PDMS silicone conditioner is

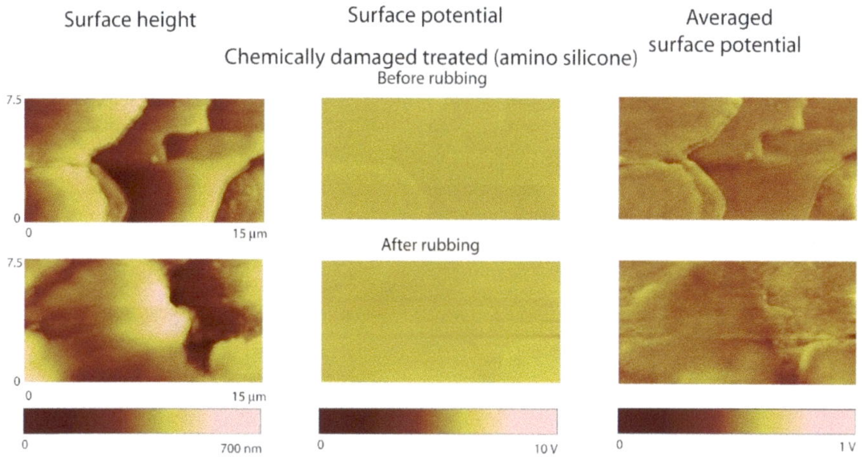

(a)

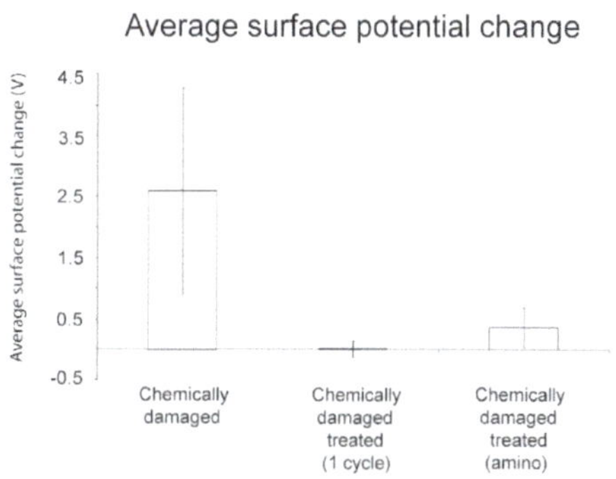

(b)

Fig. 7.2 (**a**) AFM images of chemically damaged treated (amino) hair both before and after rubbing with latex; (**b**) bar chart showing average surface potential change after rubbing with latex (Lodge and Bhushan, 2007a)

believed to only physically attach to the hair surface, and it remains mobile. Therefore, rubbing with the finger cot is likely to redistribute the conditioner, effectively spreading it evenly over the hair. However, amino silicone conditioner chemically attaches to the surface and distributes evenly to begin with. Moreover, amino silicone conditioner remains evenly distributed due to the chemical attachment. A method to rub the hair over a smaller area may provide more insight into this behavior and may elucidate the effect of conditioner distribution on surface potential properties. However, the amino silicone conditioner still shows a significant decrease in charge from the untreated chemically damaged sample.

Figure 7.3 shows mechanically damaged hair samples. This figure shows results more similar to virgin hair than to chemically damaged hair. This is likely due to the

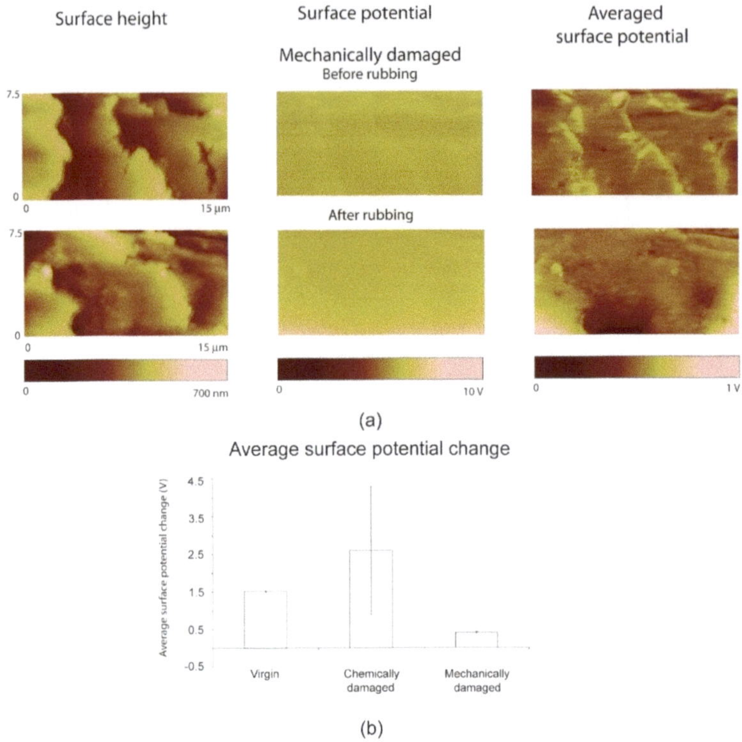

Fig. 7.3 (**a**) AFM images of mechanically damaged hair both before and after rubbing with latex; (**b**) bar chart showing average surface potential change after rubbing with latex (Lodge and Bhushan, 2007a)

fact that the top lipid layer of hair is not removed and that most of the mechanical damage occurs at scale edges. Chemical damage occurs over the entire surface of the hair. For this reason it is likely that mechanically damaged hair will behave more like virgin hair than chemically damaged hair. Because the rubbing procedure is not strictly controlled in terms of location of contact, it is difficult to distinguish a difference between the topmost layer of the hair and the portions which have been mechanically damaged. However, this indifference between these regions (which generally occur on the surface away from scale edges, and near scale edges, respectively) may also be due to the fact that physical wear does not affect surface potential in insulating samples, as discussed previously.

7.2 Effect of External Voltage and Humidity on Surface Potential

Samples were mounted in conductive silver paint and an external power supply was used to develop charge (Lodge and Bhushan, 2007b). To determine environmental effects on surface charge, relative humidity was controlled and varied.

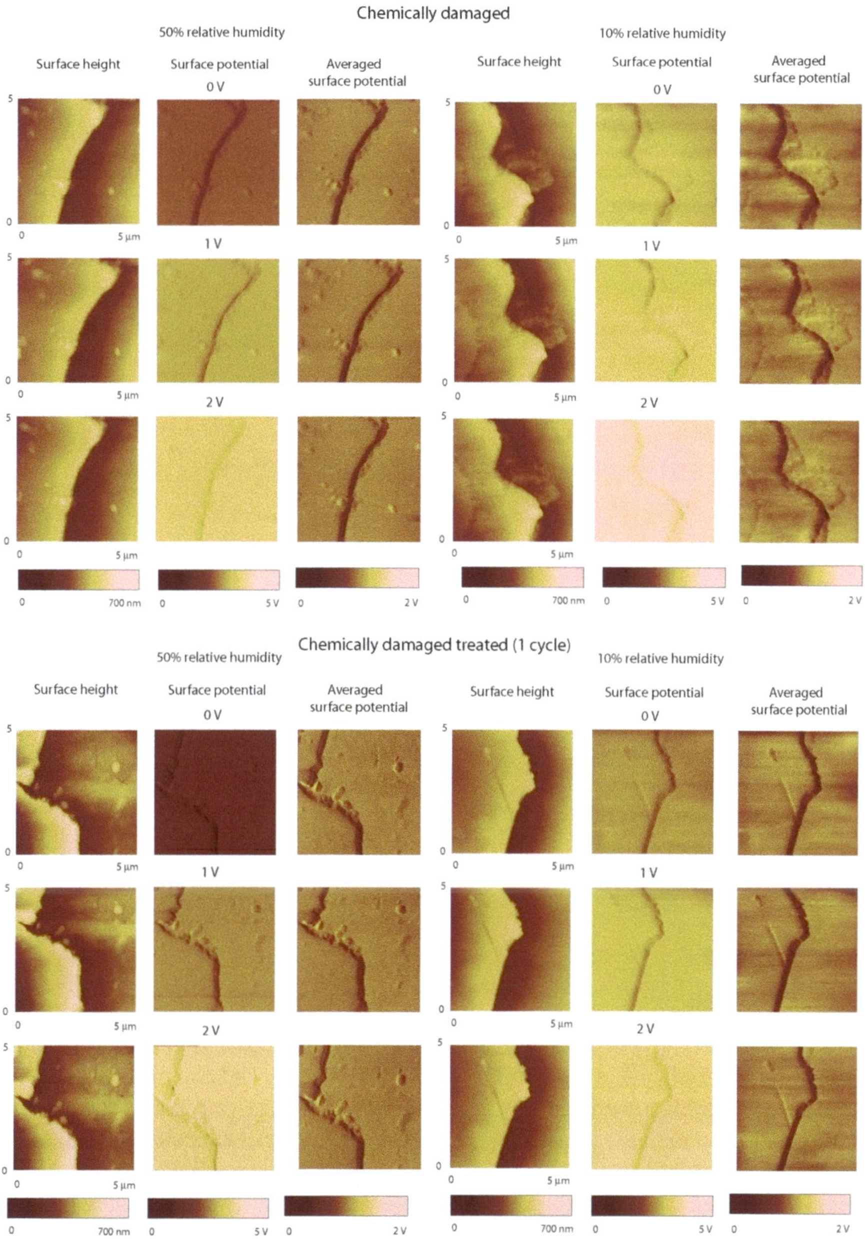

Fig. 7.4 (**a**) AFM images showing surface height and surface potential of chemically damaged and chemically damaged treated (1 cycle conditioner) hair samples at both 50 and 10% relative humidity; (**b**) bar charts showing average surface potential change at both 50 and 10% relative humidities of virgin, virgin treated, chemically damaged, chemically damaged treated (1 cycle conditioner), and chemically damaged treated (3 cycles conditioner) hair samples (Lodge and Bhushan, 2007b)

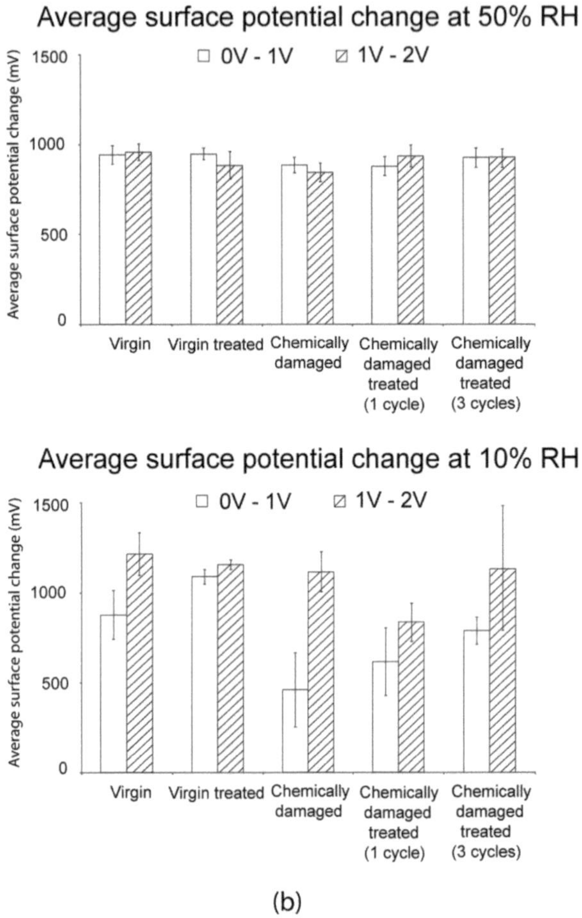

Fig. 7.4 (continued)

To study the effects of external voltage and humidity, virgin, virgin treated, chemically damaged, chemically damaged treated (one cycle), and chemically damaged treated (three cycles) were studied at an external voltage of 0, 1, and 2 V at 50 and 10% relative humidities. Representative AFM images for chemically damaged and chemically damaged treated (one cycle of conditioner) hair samples are shown in Fig. 7.4a. It should be noted that the contrast seen around cuticle scale edges is likely a result of the topography effect (Lodge and Bhushan, 2007b). A key point shown by these images is the existence of areas of trapped charge in the low-humidity samples. This is not seen in the samples measured in ambient. These areas of trapped charge are seen as bright areas on the sample in the "averaged surface potential" images. In other words, there is more contrast in the averaged surface potential maps under low humidity (neglecting any contrast seen around the scale edge) than is seen for the same samples under ambient humidity. It is important to clarify that the "bright areas" are only in reference to the increase in contrast in the surface potential map (averaged surface potential) away from the scale edges. This is only considered for the "0 V" samples, where added external charge does not affect the surface potential of the sample. This observation indicates that water vapor in the air contributes significantly to the mobility of surface charges on the hair. This result has been previously reported for macro-studies on surface charge of hair (Jachowicz et al., 1985; Lunn and Evans, 1977; Mills et al., 1956). Trapped charges are most pronounced in the untreated samples. This suggests that conditioner treatment has a similar effect as water vapor. Even under very low humidity conditions, conditioner treatment increases the mobility of surface charges, dissipating trapped charges. Figure 7.4b shows bar charts of the average surface potential change. Error bars indicate ± one standard deviation. It can be observed in these figures that all samples exhibit very similar values in the 50% relative humidity scenario. This again indicates that the water vapor in the air plays a significant role in the surface charge of the hair and also in the mobility of charge. For the 50% case it is shown that the potential changes nearly 1 V for every 1 V change in applied potential. However, this is not the case for the 10% relative humidity situation. In this case, a higher 0–1 V value indicates more charge mobility and the ability to dissipate charge, whereas

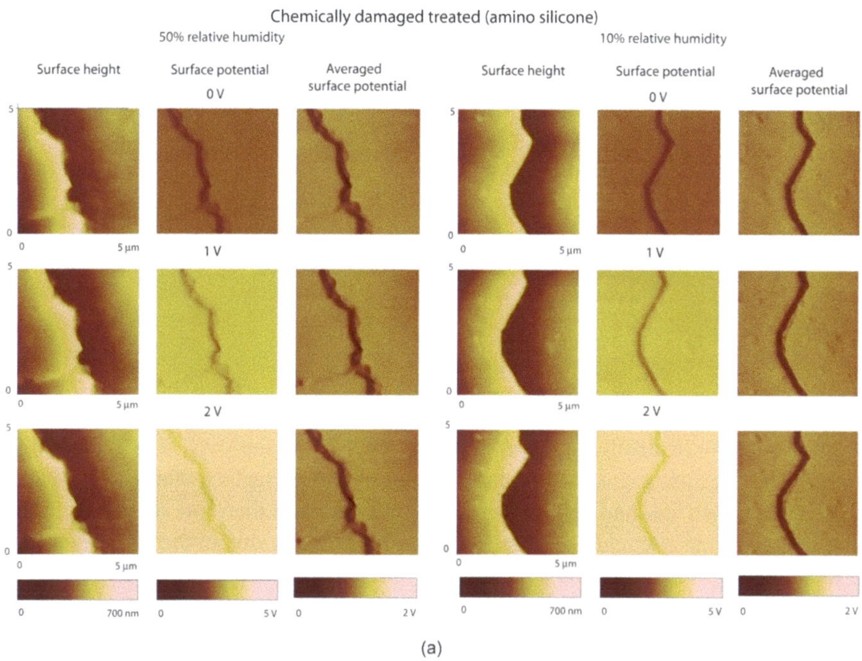

Fig. 7.5 (**a**) AFM images showing surface height and surface potential of chemically damaged treated (amino silicone) hair samples at both 50 and 10% relative humidities; (**b**) bar charts showing average surface potential change at both 50 and 10% relative humidities for chemically damaged, chemically damaged treated (1 cycle conditioner), and chemically damaged treated (amino silicone) hair samples (Lodge and Bhushan, 2007b)

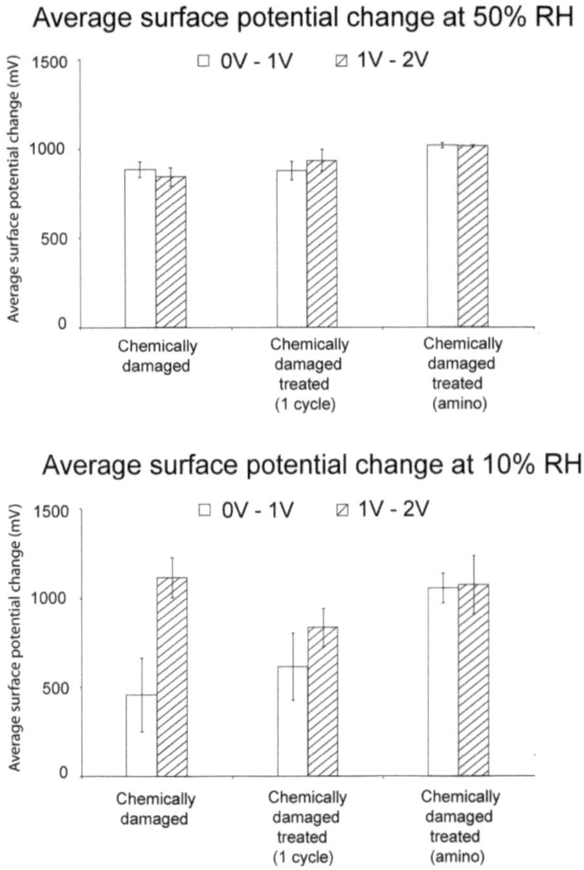

(b)

Fig. 7.5 (continued)

a lower value indicates less charge mobility. This being the case, it is evident that conditioner treatment greatly increases the mobility of surface charges. It is also shown that virgin hair has a better charge mobility and can therefore dissipate charge more readily than chemically damaged hair. This is likely due to the lipid layer that is intact in virgin hair, but has been removed in damaged hair.

Figure 7.5a shows AFM images of chemically damaged hair treated with amino silicone conditioner. Here it is observed that the surface of the hair is more or less an equipotential surface, similar to the chemically damaged treated (three cycles) sample. This suggests that the amino silicone conditioner coats the surface fully with one cycle, whereas commercial conditioner does not coat fully until several cycles. Figure 7.5b shows this behavior as well. Amino silicone conditioner shows potential changes in 10% relative humidity that are almost identical to those measured in

50% relative humidity. This indicates that the conditioner is evenly spread over the surface of the hair, which has been previously reported (LaTorre and Bhushan, 2006; Lodge and Bhushan, 2006a). It also indicates that amino silicone conditioner greatly increases charge mobility. Because commercial conditioner is unevenly distributed on the hair surface, the data for the commercially treated samples exhibit relatively high standard deviations in most cases. Consequently, different areas on the hair surface may contain vastly different amounts of conditioner on the surface. For the amino silicone case, however, different areas of the hair seem to have similar amounts of conditioner as evidenced by the lower standard deviation. The data then suggest that amino silicone conditioner is the superior conditioner in terms of charge dissipation.

7.3 Effect of Rubbing Load on Nanoscale Charging Characteristics

For nanoscale rubbing load studies, a Co/Cr-coated conducting Si tapping mode tip (MESP, Veeco Probes) with cantilever spring constant 2.8 N/m was used for both rubbing and surface potential measurement (Seshadri and Bhushan, 2008c). In order to charge a control area on the hair surface, the tip with zero voltage bias was first engaged onto the surface in contact mode and scanned once over a 40 μm × 20 μm area, at the lowest possible load (lesser than 100 nN). Next, a 10 μm × 10 μm control area was selected for rubbing. By adjusting the contact mode setpoint and looking at the force calibration plot, the rubbing normal load was set at the required value. Simultaneously, in order to produce a background charge repeatedly on the hair surface to overcome background noise, the DC voltage bias of 6V was applied to the tip, internally, through the controller. The voltage-biased tip was then rastered once over this 10 μm × 10 μm area, at the increased load. This led to charging of the hair surface. The rubbing normal load was successively increased in steps of 100 nN, from 450 to 750 nN. Macroscale experiments on combing bundles of hair a few mm long have estimated combing load to be on the order of 100 mN (Kamath and Weigman, 1986). However, controlling the combing load on the macroscale is extremely difficult, as it depends on various factors like interfiber friction and entangling. The load applied to the probe in this study was chosen such that charging occurred, but there was no noticeable damage to the tip or hair surface. This ensured repeatable charging and consistent values of measured charge, as compared to macroscale techniques.

7.3.1 *Virgin Hair*

Figure 7.6 shows the change in potential after charge deposition in virgin hair at two normal loads. The first column is the topography of the control area. The second column is the surface potential map of the same area before rubbing. The third

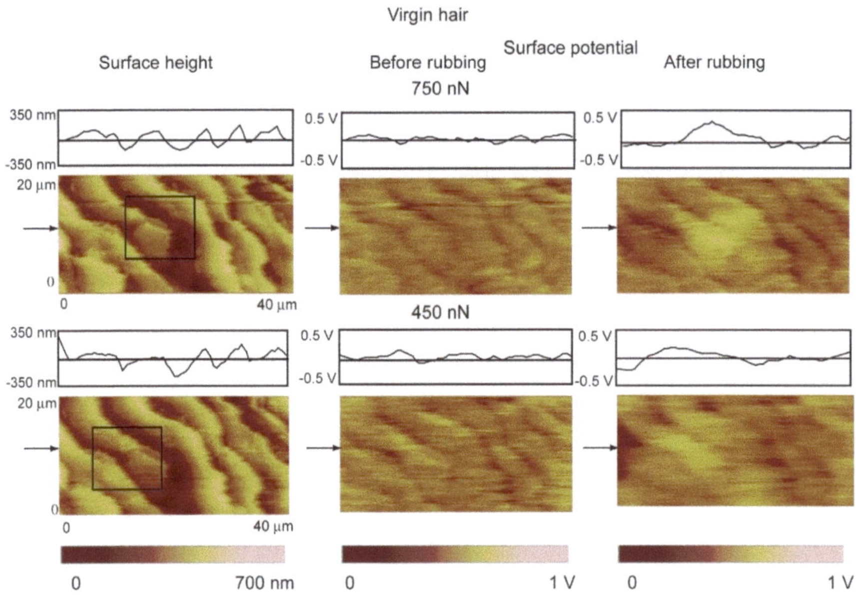

Fig. 7.6 AFM topography images and KPM surface potential maps of virgin hair before and after rubbing with voltage-biased tip at two different normal loads. The area where rubbing was performed is *marked* on the topography image (Seshadri and Bhushan, 2008c)

column is the surface potential map of the control area after rubbing. In the surface potential maps of hair, a contrast or increase in potential around the cuticle edges is generally observed. This is thought to be an artifact of the KPM technique (Lodge and Bhushan, 2007a, b) and hence will not be discussed further. By comparing the before and after rubbing surface potential maps in Fig. 7.6, it is seen that a marked increase in surface potential occurs over the marked area, where the surface was rubbed with a voltage-biased tip, even at the lowest load.

The tendency of hair to get charged while rubbing is partially due to its electret properties. Electrets are materials capable of storing charge and polarization. This makes them susceptible to getting charged and retaining that charge during rubbing and triboelectric contact. Human hair is a natural bio-electret, and similar to other polymer electrets, displays piezoelectric and pyroelectric properties (Mascarenhas, 1980; Jachowicz et al., 1984; Feughelman et al., 2003; Fukada, 1992). The natural electret properties of hair are thought to arise from the polarity of the peptide bond in the keratin molecules, and also from their synergistic arrangement in the α-helix structuring of the cortical cells, which adds up individual molecule polarities (Mascarenhas, 1980).

Figure 7.6 also shows surface potential images at two loads, and Fig. 7.7 displays a quantitative bar chart of the change in surface potential at four different rubbing normal loads. The values displayed in the bar chart are averages over the whole charged/worn area, for measurements on at least five different samples of each type of hair tested, at each load step. For virgin hair, compared to other hair samples

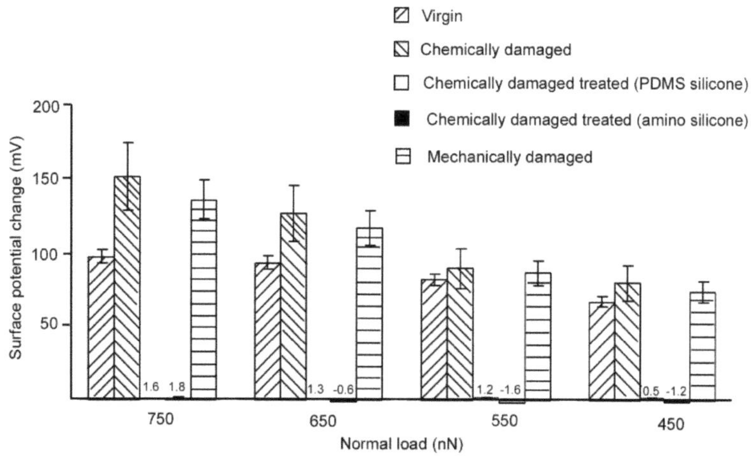

Fig. 7.7 Bar chart showing average surface potential change and its dependence on normal load for all the hair samples studied (Seshadri and Bhushan, 2008c)

on the bar chart, it is seen that the surface potential change does not vary much with increasing rubbing load. It has been shown that the friction force in virgin hair shows very little increase with normal rubbing load, compared to damaged hair. This is attributed to the intact lipid or 18 MEA layer on virgin hair (LaTorre and Bhushan, 2005a). This layer contributes significantly to the lubricity of hair surface. Charge developed is known to increase with increasing friction between hair and rubbing element (Lunn and Evans, 1977). As the friction does not increase much with normal load, the surface potential change or the charge developed also remains more or less constant.

7.3.2 Effect of Damage

Figure 7.8a shows topography and surface potential maps of chemically damaged hair, before and after rubbing. Figure 7.7 displays the corresponding quantitative values at four different loads. It is observed that at lower loads the surface potential change in chemically damaged hair is comparable to virgin hair, while at higher loads it is much more than virgin hair. In chemically damaged hair, the naturally present lipid layer is stripped away by the bleach and coloring agents (LaTorre and Bhushan, 2006; LaTorre and Bhushan, 2006). This means that there is higher friction between the tip and the surface. It has also been shown that friction increases much more rapidly with normal load in damaged hair, as compared to virgin hair (LaTorre and Bhushan, 2005a). This leads to more charge generation and the increase in the surface potential change at higher loads.

Figure 7.9 displays the topography and potential maps for mechanically damaged hair; Fig. 7.7 shows the quantitative values. Mechanically damaged hair shows surface potential values that are in between virgin hair and chemically damaged hair.

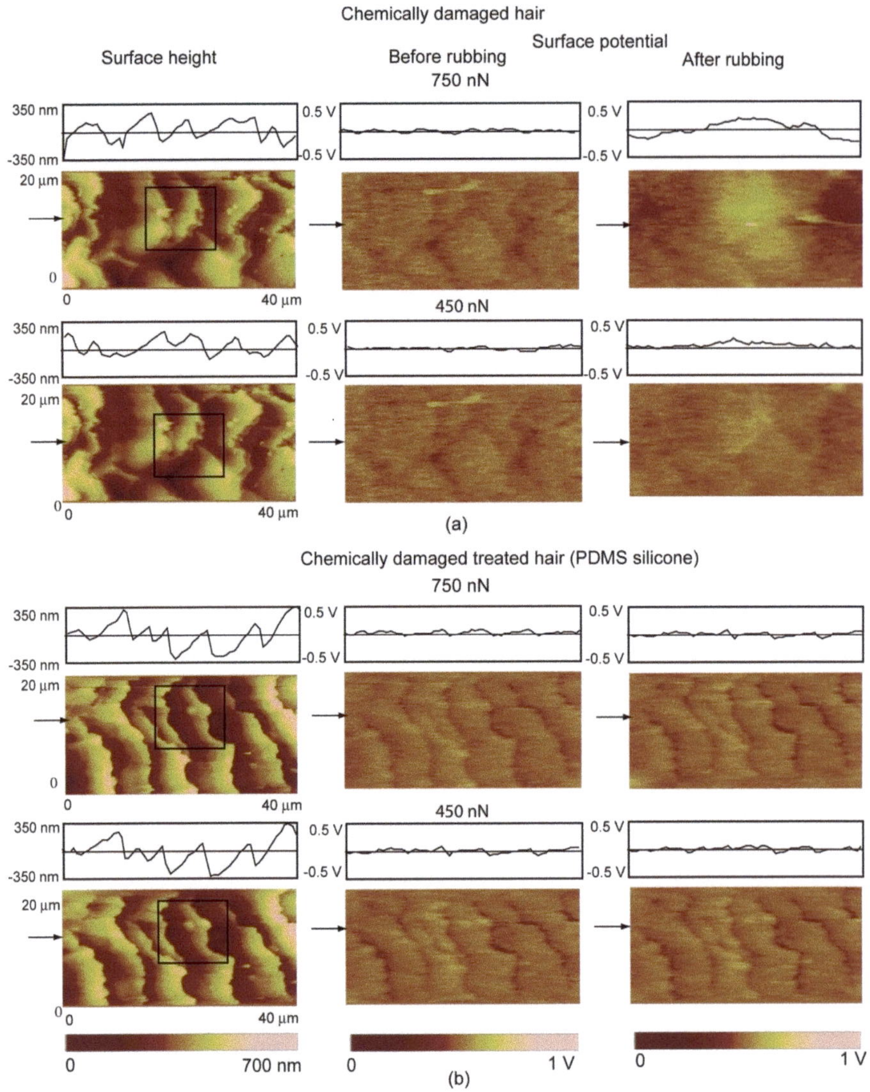

Fig. 7.8 Topography images and surface potential maps before and after rubbing with voltage-biased tip at two different normal loads for (**a**) chemically damaged hair; (**b**) chemically damaged hair treated with PDMS silicone conditioner; (**c**) chemically damaged hair treated with amino silicone conditioner. The area where rubbing was performed is *marked* on the topography image (Seshadri and Bhushan, 2008c)

The stripping of the lipid layer is complete in chemically damaged hair, because of the uniform nature of chemical damage. However, with mechanically damaged hair, it might be possible that the lipid layer is only partially stripped off, wherever the comb has rubbed into the fiber. Hence, the friction and its variation with normal load is probably not as much as chemically damaged hair. However, it is still found to be higher than virgin hair.

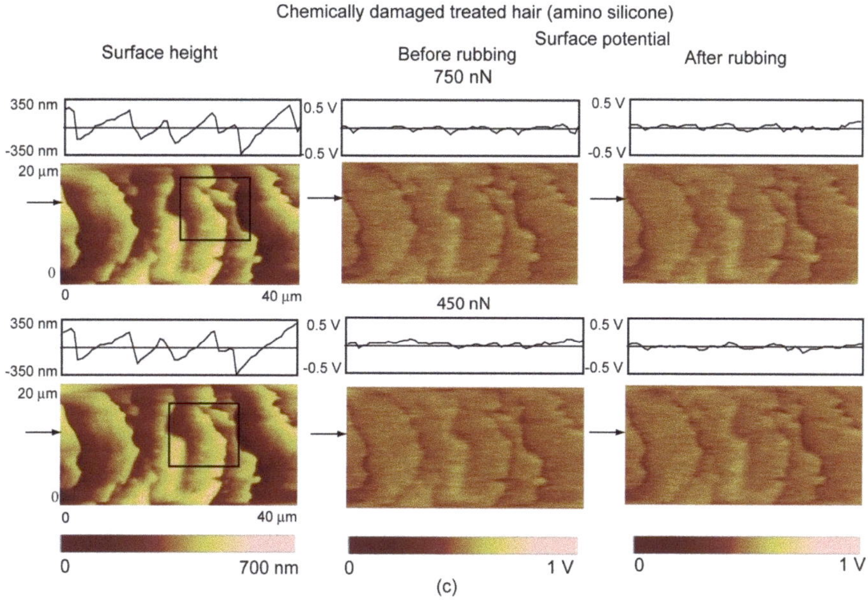

Fig. 7.8 (continued)

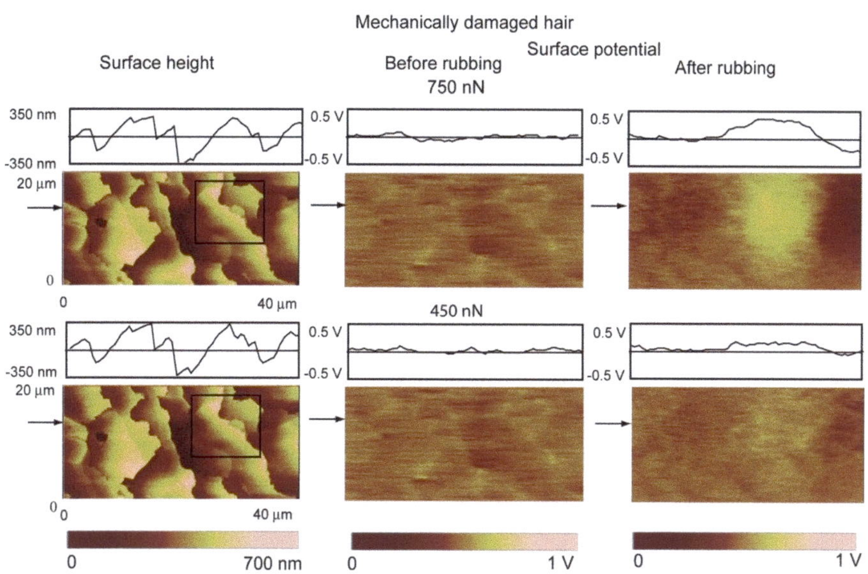

Fig. 7.9 Topography images and surface potential maps before and after rubbing with voltage-biased tip for mechanically damaged hair. The area where rubbing was performed is *marked* on the topography image (Seshadri and Bhushan, 2008c)

7.3.3 Effect of Conditioner Treatments

Figure 7.8b and c shows surface potential maps for chemically damaged conditioner-treated hair samples. Two types of conditioners were studied – PDMS silicone based conditioner and a newer amino silicone based conditioner. Figure 7.7 again displays quantitative values of surface potential change.

It is clearly seen that both conditioners dramatically reduce the surface potential change and charge deposition on the hair. There is almost no surface potential change due to rubbing on conditioner-treated hair. Conditioners are known to possess strong antistatic properties. There are two possible mechanisms by which conditioners could reduce charge on hair (Lunn and Evans, 1977). The first is the lubricating effect of the conditioners. Conditioners are basically silicone oils that lubricate hair surface. They replenish the 18 MEA layer stripped by damage. This lower friction leads to the lower charging. However, the second and probably more dominant effect is the increase in surface conductivity. Hersh et al. (1981) have reported that conditioners increase the conductivity of hair fibers. The resistance per unit length of hair decreased from 10^{18} to 10^{17} ohm/cm after conditioner treatment. So, whatever charge is deposited is quickly dissipated by the conducting conditioner layer.

Both amino silicone and PDMS silicone treated hair samples show comparable surface potential changes. PDMS silicone physically attaches to the surface while amino silicone is chemically bonded, and hence more uniformly covers the surface. However, not much difference is seen between the two conditioners' antistatic effects. This could be because surface conductivity is probably the dominant antistatic mechanism. Even a non-uniform conditioner layer significantly increases the conductivity. Also, rubbing with the AFM tip may actually be redistributing the PDMS conditioner more uniformly over the surface, especially at higher loads. Still, both the amino silicone and PDMS silicone conditioners show a significant reduction of potential compared to damaged or even virgin hair.

7.4 Summary

Kelvin probe microscopy is a powerful tool when used to study the electrostatic properties of hair. Physical wear alone does not cause a measurable electrostatic charge on hair because hair is non-conductive. Rubbing with a latex finger cot can build up electrostatic charge on hair. Any electrostatic charge on hair may dissipate depending upon the condition of the hair. The natural lipid layer on virgin hair and conditioner treatment may increase the rate of dissipation. Untreated chemically damaged hair, which has no lipid layer, seems to retain charge the longest. Conditioner treatment significantly decreases the charge built on the hair surface. Amino silicone conditioner coats the hair surface uniformly and provides higher charge mobility and less likelihood of trapped charges than the commercial conditioner. Relative humidity plays a significant role in the behavior of surface charges on hair.

A low relative humidity decreases charge mobility, thereby increasing the likelihood of charges trapped on the surface.

Rubbing with a voltage-biased AFM tip is an effective method to study charging characteristics of human hair on a nanoscale in a controlled and systematic way. Under such conditions, virgin hair shows a marked potential difference in the control area on rubbing. However, the potential difference remains almost constant with increasing rubbing loads. Chemically damaged hair shows a higher potential change than virgin hair and also greater surface potential change with increasing normal loads. Mechanically damaged hair shows a potential change between chemically damaged and virgin hair, because the lubricating lipid layer is only partially stripped away. Both PDMS silicone and amino silicone conditioners significantly reduce potential change. This is because they increase the surface conductivity and lubricity drastically.

Appendix A
Shampoo and Conditioner Treatment Procedure

This section outlines the steps involved in washing hair swatches with shampoo and/or conditioner.

A.1 Shampoo Treatments

Shampoo treatments consisted of applying a commercial shampoo evenly down a hair swatch with a syringe. The hair was lathered for 30 s, rinsed with tap water for 30 s, and the process was repeated. The amount of shampoo used for each hair swatch was 0.1 cm^3 of shampoo per gram of hair. Swatches were hung to dry in an environmentally controlled laboratory and then wrapped in aluminum foil.

A.2 Conditioner Treatments

A commercial conditioner was applied, 0.1 cm^3 of conditioner per gram of hair. The conditioner was applied in a downward direction (scalp to tip) thoroughly throughout the hair swatch for 30 s. The swatch was then rinsed thoroughly for 30 s. Swatches were hung to dry in an environmentally controlled laboratory and then wrapped in aluminum foil.

B. Bhushan, *Biophysics of Human Hair*, Biological and Medical Physics, Biomedical Engineering, DOI 10.1007/978-3-642-15901-5,

Appendix B
Conditioner Thickness Approximation

We consider a cylindrical hair fiber of diameter $D = 50\ \mu\text{m}$ (radius $R = 25\ \mu\text{m}$). For conditioner thickness calculations, the following assumptions are made: (a) hair and the material being added have the same density; (b) the coating of material is uniform on the hair surface; (c) the cross-sectional area of a hair fiber remains constant along the longitudinal axis of the fiber (i.e., from root to tip); the hair fiber is perfectly cylindrical (circular cross section); and (d) the deposited conditioner remains bonded to the cuticle surface (no absorption into the cuticle layer).

The cross-sectional area of an untreated hair fiber is initially calculated. By adding a specified amount of conditioner, this area will increase and cause the radius of the treated hair fiber to increase. This increase in the radius of the treated hair will be equivalent to the thickness of the conditioner layer. The original cross-sectional area A_c of the hair fiber is

$$A_c = \pi R^2 = \pi(25\mu\text{m})^2 = 1963.4954\,\mu\text{m}^2$$

Adding 200 ppm of material to the surface (which is comparable to the amount that commercial conditioners typically deposit) will cause an increase in volume (for a unit fiber length) by 200 ppm, or by 0.0002. Thus, the cross-sectional area $A_{c,\text{conditioner}}$ of the treated hair will increase by the same amount to

$$A_{c,\text{conditioner}} = 1.0002\,A_c = 1963.888\,\mu\text{m}^2$$

which results in a new radius $R_{\text{conditioner}}$

$$R_{\text{Conditioner}} = \sqrt{A_{c,\text{Conditioner}}/\pi} = 25.0025\,\mu\text{m}$$

Therefore, subtracting the original radius from the radius after treatment increases the thickness of the hair by 0.0025 μm, or 2.5 nm.

It is important to note that the approximation of the conditioner thickness as 2.5 nm was determined for a particular hair diameter and material deposition amount (with the hair and material having equal densities). Although these are generally realistic approximations, hair diameter often varies by a factor of 2 and the deposition level can vary up to an order of magnitude. The conditioner layer has been shown in previous work to be non-uniform as well. Thus, actual conditioner thickness can deviate significantly from this number.

References

Attard P, Miklavcic SJ (2001) Effective spring constant of bubbles and droplets. Langmuir 17:8217–8223

Barnes HA, Roberts GP (2000) The non-linear viscoelastic behaviour of human hair at moderate extensions. Inter J Cosmet Sci 22:259–264

Benz M, Chen NH, Israelachvili J (2004) Lubrication and wear properties of grafted polyelectrolytes, hyaluronan and hylan, measured in the surface forces apparatus. J Biomed Mater Res A 71A:6–15

Beyak R, Meyer CF, Kass GS (1969) Elasticity and tensile properties of human hair I – single fiber test method. J Soc Cosmet Chem 20:615

Bhatt D, Newman J, Radke CJ (2001) Equilibrium force isotherms of a deformable bubble/drop interacting with a solid particle across a thin liquid film. Langmuir 17:116–130

Bhushan B (1984) Influence of test parameters in the measurement of the coefficient of friction of magnetic tapes. Wear 93:81–89

Bhushan B (1999a) Principles and applications of tribology. Wiley, New York, NY

Bhushan B (1999b) Handbook of micro/nanotribology. 2nd edn. CRC Press, Boca Raton, FL

Bhushan B (2002) Introduction to tribology. Wiley, New York, NY

Bhushan B (2008a) Nanoscale characterization of human hair and hair conditioners. Prog Mater Sci 53:585–710

Bhushan B (2008b) Nanotribology and nanomechanics – an introduction, 2nd edn. Springer, Heidelberg

Bhushan B, Blackman GS (1991) Atomic force microscopy of magnetic rigid disks and sliders and its applications to tribology. J Tribol Trans ASME 113:452–457

Bhushan B, Burton Z (2005) Adhesion and friction properties of polymers in microfluidic devices. Nanotechnology 16:467–478

Bhushan B, Chen N (2006) AFM studies of environmental effects on nanomechanical properties and cellular structure of human hair. Ultramicroscopy 106:755–764

Bhushan B, Dandavate C (2000) Thin-film friction and adhesion studies using atomic force microscopy. J Appl Phys 87:1201–1210

Bhushan B, Goldade AV (2000a) Measurement and analysis of surface potential change during wear of single-crystal silicon (100) at ultralow loads using Kelvin probe microscopy. Appl Surf Sci 157:373–381

Bhushan B, Goldade AV (2000b) Kelvin probe microscopy measurements of surface potential change under wear at low loads. Wear 244:104–117

Bhushan B, Kasai T (2004) A surface topography-independent friction measurement technique using torsional resonance mode in an AFM. Nanotechnology 15:923–935

Bhushan B, Li X (2003) Nanomechanical properties of solid surfaces and thin films (invited). Inter Mater Rev 48:125–164

Bhushan B, Qi J (2003) Phase contrast imaging of nanocomposites and molecularly thick lubricant films in magnetic media. Nanotechnology 14:886–895

Bhushan B, Liu H, Hsu SM (2004) Adhesion and friction studies of silicon and hydrophobic and low friction films and investigation of scale effects. ASME J Tribol 126:583–590

Bhushan B, Wei G, Haddad P (2005) Friction and wear studies of human hair and skin. Wear 259:1012–1021

Bobji MS, Bhushan B (2001a) In situ microscopic surface characterization studies of polymeric thin films during tensile deformation using atomic force microscopy. J Mater Res 16:844–855

Bobji MS, Bhushan B (2001b) Atomic force microscopy study of the microcracking of magnetic thin films under tension. Scripta Mater 44:37–42

Bolduc C, Shapiro J (2001) Hair care products: waving, straightening, conditioning, and coloring. Clin Dermatol 19:431–436

Brandup J, Immergut EH, Grulke EA (eds) (1999) Polymer handbook (4th edn). Wiley, New York, NY

Chen N, Bhushan B (2005) Morphological, nanomechanical and cellular structural characterization of human hair and conditioner distribution using torsional resonance mode in an AFM. J Microscopy 220:96–112

Chen N, Bhushan B (2006) Atomic force microscopy studies of conditioner thickness distribution and binding interactions on the hair surface. J Microscopy 221:203–215

Chen NH, Kuhl T, Tadmor R, Lin Q, Israelachvili JN (2004) Large deformations during the coalescence of fluid interfaces. Phys Rev Lett 92:Art. No.-024501

Dekoi S, Jedoi J (1988) Hair low-sulphur protein composition does not differ electrophoretically among different races. J Dermatol 15:393–396

Dekoi S, Jedoi J (1990) Amount of fibrous and matrix substances from hair of different races. J Dermatol 19:62–64

DeVecchio D, Bhushan B (1998) Use of a nanoscale Kelvin probe for detecting wear precursors. Rev Sci Instrum 69:3618–3824

P Elsner, E Berardesca, Maibach HI (ed) (1994) Bioengineering of the skin: water and the stratum corneum. CRC Press, Boca Raton, FL

Feughelman A (1997) Mechanical properties and structure of alpha-keratin fibres: wool, human hair and related fibres. University of South Wales Press, Sydney

Feughelman A, Lymanb D, Menefee E, Willis B (2003) The orientation of the α-helices in α-keratin fibres. Intl J Biol Macromol 33:149–152

Feughelman M (1982) The physical properties of alpha keratin fibers. J Soc Cosmet Chem 33:385–406

Forcada ML, Jakas MM, Grasmarti A (1991) On liquid-film thickness measurements with the atomic-force microscope. J Chem Phys 95:706–708

Fukada E (1992) Bio-electrets and bio-piezoelectricity. IEEE Trans Elec Insul 27:813–819

Ginn ME, Noyes CM, Jungermann E (1968) The contact angle of water on viable human skin. J Colloid Interface Sci 26:146–151

Gray J (2001) Hair care and hair care products. Clin Dermatol 19:227–236

Gray J (2003) The world of hair (Online, http://www.pg.com/science/haircare/hair_twh_toc.htm)

Henderson GH, Karg GM, O'Neill JJ (1978) Fractography of human hair. J Soc Cosmet Chem 29:449–467

Hersh SP, Grady PL, Bhat GR (1981) Effect of internal and external tensides on the electrical properties of plymeric surface. Pure Appl Chem 53:2123–2134

Israelachvili: JN (1992) Intermolecular & surface forces, 2nd edn. Academic, London

Jachowicz J, McMullen R (2002) Mechanical analysis of elasticity and flexibility of virgin and polymer-treated hair fiber assemblies. J Cosmet Sci 53:345–361

Jachowicz J, Wis-Surel G, Wolfram LJ (1984) Directional triboelectric effect in keratin fibers. Text Res J 54:492–495

Jachowicz J, Wis-Surel G, Garcia ML (1985) Relationship between triboelectric charging and surface modifications of human hair. J Soc Cosmet Chem 36:189–212

Jalbert C, Koberstein JT, Yilgor I, Gallagher P, Krukonis V (1993) Molecular weight dependence and end-group effects on the surface tension of poly(dimethylsiloxane). Macromolecules 26:3069–3074

Jollès P, Zahn H, Hoecker H (eds) (1997) Formation and structure of human hair. Birkhäuser Verlag, Berlin

Kamath YK, Hornby SB (1984) Mechanical and fractographic behavior of negroid hair. J Soc Cosmet Chem 35:21–43

Kamath YK, Weigman HD (1986) Measurement of combing forces. J Soc Cosmet Chem 37:111–116

Kasai T, Bhushan B, Huang L, Su CM (2004) Topography and phase imaging using the torsional resonance mode. Nanotechnology 15:731–742

LaTorre C, Bhushan B (2005a) Nanotribological characterization of human hair and skin using atomic force microscopy. Ultramicroscopy 105:155–175

LaTorre C, Bhushan B (2005b) Nanotribological effects of hair care products and environment on human hair using atomic force microscopy. J Vac Sci Technol A 23:1034–1045

LaTorre C, Bhushan B (2006) Investigation of scale effects and directionality dependence on adhesion and friction of human hair using AFM and macroscale friction test apparatus. Ultramicroscopy 106:720–734

LaTorre C, Bhushan B, Yang JZ, Torgerson PM (2006) Nanotribological effects of silicone type, silicone deposition level, and surfactant type on human hair using atomic force microscopy. J Cosmetic Sci 57:37–56

Lerebour G, Cupferman S, Cohen C, Bellon-Fontaine MN (2000) Comparison of surface free energy between reconstructed human epidermis and in situ human skin. Skin Res Technol 6:245–249

Leyden JJ, Rawlings AV (ed) (2002) Skin moisturization. Marcel Dekker, New York, NY

Li X, Bhushan B, McGinnis PB (1996) Nanoscale mechanical characterization of glass fibers. Mater Lett 29:215–220

Lindelof B, Forslind B, Hedblad MS (1988) Human hair form. Morphology revealed by light and scanning electron microscopy and computer aided three-dimensional reconstruction. Arch Dermatol 124:1359–1363

Liu H, Bhushan B (2003) Nanotribological characterization of molecularly thick lubricant films for applications to MEMS/NEMS by AFM. Ultramicroscopy 97:321–340

Lodén M, Maibach HI (ed) (2000) Dry skin and moisturizers: chemistry and function. CRC Press, Boca Raton, FL

Lodge RA, Bhushan B (2006a) Surface characterization of human hair using tapping mode atomic force microscopy and measurement of conditioner thickness distribution. J Vac Sci Technol A 24:1258–1269

Lodge RA, Bhushan B (2006b) Wetting properties of human hair By Means Of dynamic contact angle measurement. J Appl Poly Sci 102:5255–5265

Lodge RA, Bhushan B (2007a) Effect of physical wear and triboelectric interaction on surface charges measured by Kelvin probe microscopy. J Colloid Interface Sci 310:321–330

Lodge RA, Bhushan B (2007b) Surface potential measurement of human hair using Kelvin probe microscopy. J Vac Sci Technol A 25:893–902

Lunn AC, Evans RE (1977) The electrostatic properties of human hair. J Soc Cosmet Chem 28:549–569

Mark JE (1999) Polymers data handbook. Oxford University Press, Oxford

Mascarenhas S (1980)Bioelectrets: electrets in biomaterials and biopolymers. In: Sessler GM (ed) Electrets, topics in applied physics, vol **33**. Springer, Heidelberg

Menkart J, Wolfram LJ, Mao I (1984) Caucasian hair, negro hair, and wool: similarities and differences. J Soc Cosmet Chem 35:21–43

Mills CM, Ester VC, Henkin H (1956) Measurement of static charge on hair. J Soc Cosmet Chem 7:466–475

Molina R, Comelles F, Julia MR, Erra P (2001) Chemical modifications on human hair studied by means of contact angle determination. J Colloid Interface Sci 237:40–46

Nappe C, Kermici M (1989) Electrophoretic analysis of alkylated proteins of human hair from various ethnic groups. J Cosmet Chem 40:91–99

Negri AP, Cornell HJ, Rivett DE (1993) A model for the surface of keratin fibers. Textile Res J 63:109–115

Nikiforidis G, Balas C, Tsambaos D (1992) Mechanical parameters of human hair: possible applications in the diagnosis and follow-up of hair disorders. Clin Phys Physiol Meas 13:281–290

Parbhu AN, Bryson WG, Lal R (1999) Disulfide bonds in the outer layer of keratin fibers confer higher mechanical rigidity: correlative nano-indentation and elasticity measurement with an AFM. Biochemistry 38:11755–11761

Randebrook RJ (1964) Neue Erkenntnisse über den Morphologischen Aufbau des Menschlichen Haares, J Soc Cosmet Chem 15:691–706

Robbins C (1994) Chemical and physical behavior of human hair, 3rd edn. Springer, New York, NY

Robbins CR, Crawford RJ (1991) Cuticle damage and the tensile properties of human hair. J Soc Cosmet Chem 42:59–67

Reutsch SB, Weigmann HG (1996) Mechanism of tensile stress release in the keratin fiber. J Soc Cosmet Chem 47:13–26

Schott H (1971) Contact angles and wettability of human skin. J Pharm Sci 60:1893–1895

Scott GV, Robbins CR (1980) Effects of surfactant solutions on hair fiber friction. J Soc Cosmet Chem 31:179–200

Scott WW, Bhushan B (2003) Use of phase imaging in atomic force microscopy for measurement of viscoelastic contrast in polymer nanocomposites and molecularly thick lubricant films. Ultramicroscopy 97:151–169

Seshadri IP, Bhushan B (2008a) In-situ tensile deformation characterization of human hair with atomic force microscopy. Acta Mater 56:774–781

Seshadri IP, Bhushan B (2008b) Effect of ethnicity and treatments on in situ tensile response and morphological changes of human hair characterized by atomic force microscopy. Acta Mater 56:3585–3597

Seshadri IP, Bhushan B (2008c) Effect of rubbing and load on nanoscale charging characteristics of human hair characterized by AFM based Kelvin probe. J Colloid Interf Sci 325:580–587

Shai A, Maibach HI, Baran R (2001) Handbook of cosmetic skin care. Martin Dunitz, London

Smith JR, Swift JA (2002) Lamellar subcomponents of the cuticular cell membrane complex of mammalian keratin fibres show friction and hardness contrast by AFM. J Microscopy 206:182–193

Song Y, Bhushan B (2005) Quantitative extraction of in-plane surface properties using torsional resonance mode of atomic force microscopy. J Appl Phys 97:083533

Swanbeck G, Nyren J, Juhlin L (1970) Mechanical properties of hair from patients with different types of hair diseases. J Invest Dermatol 54:248–251

Swift JA (1991) Fine details on the surface of human hair. Int J Cosmetic Sci 13:143–159

Swift JA (1997) Morphology and histochemistry of human hair. In: Jolles P, Zahn H, Hoecker H (eds) Formation and structure of human hair, Birkhäuser Verlag, Berlin, pp 149–175

Swift JA (1999) The mechanics of fracture of human hair. Inter J Cosmet Sci 21:227–239

Swift JA (2000) The cuticle controls bending stiffness of hair. J Cosmet Sci 51:37–38

Swift JA, Bews B (1974) The chemistry of human hair cuticle-I: A new method for the physical isolation of the cuticle. J Soc Cosmet Chem 25:13–22

Syed AN, Kuhajda A, Ayoub H, Ahmad K, Frank EM (1995) African-American hair: its physical properties and differences relative to Caucasian hair. Cosmet Toilet Mag 110:39–48

Tambe NS, Bhushan B (2004) In situ study of nano-cracking in multilayered magnetic tapes under monotonic and fatigue loading using an AFM. Ultramicroscopy 100:359–373

Tao Z, Bhushan B (2006a) Surface modification of AFM Si_3N_4 probes for adhesion/friction reduction and imaging improvement. ASME J Tribol 128:865–875

Tao Z, Bhushan B (2006b) Wetting properties of AFM probes By Means Of contact angle measurements. J Phys Appl Phys 39:3858–3862

Thibaut S, Bernard BA (2005) The biology of hair shape. Int J Dermatol 44:S

Thibaut S, Gailard O, Bouhanna P, Cannell DW, Bernard BA (2005) Human hair shaped is programmed from the bulb. Br J Dermatol 152:632–638

Wei G, Bhushan B (2006) Nanotribological and nanomechanical characterization of human hair using a nanoscratch technique. Ultramicroscopy 106:742–754

Wei G, Bhushan B, Torgerson PM (2005) Nanomechanical characterization of human hair using nanoindentation and SEM. Ultramicroscopy 105:155–175

Wortmann FJ, Zahn H (1994) The stress/strain curve of α-keratin fibers and the structure of the intermediate filament. Text Res J 64:737–743

Yanazawa H (1984) Adhesion model and experimental-verification for polymer SIO2 system. Colloids Surf 9:133–145

Zviak C (ed) (1986) The science of hair care. Marcel Dekker, New York, NY

Subject Index

Note: The letters 'f' and 't' following the locators refer to figures and tables respectively.

D

Biography

Dr. Bharat Bhushan received an MS in mechanical engineering from the Massachusetts Institute of Technology in 1971, an MS in mechanics and a PhD in mechanical engineering from the University of Colorado at Boulder in 1973 and 1976, respectively; an MBA from Rensselaer Polytechnic Institute at Troy, NY, in 1980; Doctor Technicae from the University of Trondheim at Trondheim, Norway, in 1990; a Doctor of Technical Sciences from the Warsaw University of Technology at Warsaw, Poland, in 1996; and Doctor Honouris Causa from the National Academy of Sciences at Gomel, Belarus, in 2000. He is a registered professional engineer. He is presently an Ohio Eminent Scholar and The Howard D. Winbigler Professor in the College of Engineering, and the Director of the Nanoprobe Laboratory for Bio- & Nanotechnology and Biomimetics (NLB^2) at the Ohio State University, Columbus, Ohio. His research interests include fundamental studies with a focus on scanning probe techniques in the interdisciplinary areas of bio/nanotribology, bio/nanomechanics, and bio/nanomaterials characterization and applications to bio/nanotechnology and biomimetics. He is an internationally recognized expert of bio/nanotribology and bio/nanomechanics using scanning probe microscopy and is one of the most prolific authors. He is considered by some a pioneer of the tribology and mechanics of magnetic storage devices. He has authored 6 scientific books, more than 90 handbook chapters, more than 700 scientific papers (h-index – 45+; ISI Highly Cited in Materials Science, since 2007), and more than 60 technical reports; edited more than 45 books; and holds 17 US and foreign patents. He is co-editor of Springer NanoScience and Technology Series and co-editor of Microsystem Technologies. He has given more than 400 invited presentations in the six continents and more than 140 keynote/plenary addresses at major international conferences.

Dr. Bhushan is an accomplished organizer. He organized the first symposium on Tribology and Mechanics of Magnetic Storage Systems in 1984 and the first international symposium on Advances in Information Storage Systems in 1990, both of which are now held annually. He is the founder of an ASME Information Storage and Processing Systems Division founded in 1993 and served as the founding chair during 1993–1998. His biography has been listed in over two dozen Who's Who books including *Who's Who in the World* and he has received more than two dozen awards for his contributions to science and technology from professional societies, industry, and US government agencies. He is also the recipient of various international fellowships including the Alexander von Humboldt Research Prize for Senior Scientists, Max Planck Foundation Research Award for Outstanding Foreign Scientists, and the Fulbright Senior Scholar Award. He is a foreign member of the International Academy of Engineering (Russia), Byelorussian Academy of Engineering and Technology, and the Academy of Triboengineering of Ukraine, an honorary member of the Society of Tribologists of Belarus, a fellow of ASME, IEEE, STLE, and the New York Academy of Sciences, and a member of ASEE, Sigma Xi and Tau Beta Pi.

Dr. Bhushan has previously worked for the R&D Division of Mechanical Technology Inc., Latham, NY; the Technology Services Division of SKF Industries Inc., King of Prussia, PA; the General Products Division Laboratory of IBM Corporation, Tucson, AZ; and the Almaden Research Center of IBM Corporation, San Jose, CA. He has held visiting professor appointments at University of California at Berkeley, University of Cambridge, UK, Technical University Vienna, Austria, University of Paris, Orsay, ETH Zurich, and EPFL Lausanne.

MIX
Papier aus verantwortungsvollen Quellen
Paper from responsible sources
FSC® C105338

If you have any concerns about our products,
you can contact us on
ProductSafety@springernature.com

In case Publisher is established outside the EU,
the EU authorized representative is:
Springer Nature Customer Service Center GmbH
Europaplatz 3, 69115 Heidelberg, Germany

Printed by Libri Plureos GmbH
in Hamburg, Germany